Sustainable Energy

*Energy, Climate and the Environment Series*

Series Editor: David Elliott, Emeritus Professor of Technology, Open University, UK

*Titles include*:

David Elliott (*editor*)
NUCLEAR OR NOT?
Does Nuclear Power Have a Place in a Sustainable Future?

David Elliott (*editor*)
SUSTAINABLE ENERGY
Opportunities and Limitations

Horace Herring and Steve Sorrell (*editors*)
ENERGY EFFICIENCY AND SUSTAINABLE CONSUMPTION
The Rebound Effect

Matti Kojo and Tapio Litmanen (*editors*)
THE RENEWAL OF NUCLEAR POWER IN FINLAND

Catherine Mitchell
THE POLITICAL ECONOMY OF SUSTAINABLE ENERGY

Ivan Scrase and Gordon MacKerron (*editors*)
ENERGY FOR THE FUTURE
A New Agenda

Gill Seyfang
SUSTAINABLE CONSUMPTION, COMMUNITY ACTION AND THE NEW ECONOMICS
Seeds of Change

Joseph Szarka
WIND POWER IN EUROPE
Politics, Business and Society

**Energy, Climate and the Environment**
**Series Standing Order ISBN 978-0-230-00800-7 (hb) 978-0-230-22150-5 (pb)**

You can receive future titles in this series as they are published by placing a standing order. Please contact your bookseller or, in case of difficulty, write to us at the address below with your name and address, the title of the series and the ISBN quoted above.

Customer Services Department, Macmillan Distribution Ltd, Houndmills, Basingstoke, Hampshire RG21 6XS, England

# Contents

# List of Figures

# List of Tables

# Series Editor Preface

# Energy, Climate and the Environment

Concerns about the potential environmental, social and economic impacts of climate change have led to a major international debate over what could and should be done to reduce emissions of greenhouse gases, which are claimed to be the main cause. There is still a scientific debate over the likely scale of climate change, and the complex interactions between human activities and climate systems, but, in the Words of no less than the Governor of California, Arnold Schwarzenegger, *'I say the debate is over. We know the science, we see the threat, and the time for action is now.'*

Whatever we now do, there will have to be a lot of social and economic adaptation to climate change – preparing for increased flooding and other climate related problems. However, the more fundamental response is to try to reduce or avoid the human activities that are seen as causing climate change. That means, primarily, trying to reduce or eliminate emission of greenhouse gasses from the combustion of fossil fuels in vehicles and power stations. Given that around 80% of the energy used in the world at present comes from these sources, this will be a major technological, economic and political undertaking. It will involve reducing demand for energy (via lifestyle choice changes), producing and using whatever energy we still need more efficiently (getting more from less), and supplying the reduced amount of energy from non-fossil sources (basically switching over to renewables and/or nuclear power).

Each of these options opens up a range of social, economic and environmental issues. Industrial society and modem consumer cultures have been based on the ever-expanding use of fossil fuels, so the changes required will inevitably be challenging. Perhaps equally inevitable are disagreements and conflicts over the merits and demerits of the various options and in relation to strategies and policies for pursuing them. These conflicts and associated debates sometimes concern technical issues, but there are usually also underlying political and ideological commitments and agendas which shape, or at least colour, the ostensibly technical debates. In particular, at times, technical assertions can be used to buttress specific policy frameworks in ways which subsequently prove to be flawed.

The aim of this series is to provide texts which lay out the technical, environmental and political issues relating to the various proposed policies for responding to climate change. The focus is not primarily on the science of climate change, or on the technological detail, although there will be accounts of the state of the art, to aid assessment of the viability of the various options. However, the main focus is the policy conflicts over which strategy to pursue. The series adopts a critical approach and attempts to identify flaws in emerging policies, propositions and assertions. In particular, it seeks to illuminate counter-intuitive assessments, conclusions and new perspectives. The aim is not simply to map the debates, but to explore their structure, their underlying assumptions and their limitations. Texts are incisive and authoritative sources of critical analysis and commentary, indicating clearly the divergent views that have emerged and also identifying the shortcomings of these views. However the books do not simply provide an overview, they also offer policy prescriptions.

The present volume attempts to review the potential for moving to a sustainable energy future by looking critically at some of the key areas of technological development, and in particular at some of the misconceptions that have grown up about what can and cannot be achieved by technology. There are certainly contested views concerning many aspects of sustainable energy. For example, while some people see renewable energy as the only long-term energy supply option, others see it as too diffuse, variable and expensive. This book looks at some of these issues – for example intermittency. However, although there may be 'technical' answers to some of these uncertainties, and new technologies are emerging which may change to situation, it is argued that the assessment of the viability of a sustainable energy future does not just depend on the merits or otherwise of specific technologies. It also depends on what sort of overall energy system we are talking about and what sort of pattern of energy use is assumed. For example, what is possible in a system based on centralised generation and ever increasing consumer demand for power, may be very different from what is possible in a decentralised system with a more sustainable approach to consumption.

Clearly then a simple technical treatise is not sufficient. We also need to look at what sort of future we are aiming for, and then at how we might move towards it in terms, for example, of providing the right kind of social and economic incentives. So while the debate over the technological options is very much a major focus, this book also looks at the perhaps even more contentious issue of how we might change patterns of energy use by the adoption of changed lifestyles and economic priorities.

# Acknowledgements

Thanks are due to Sally Boyle from the Open University Department of Design and Innovation, for her work on the graphics.

# Authors

**David Elliott** is Emeritus Professor of Technology Policy at the Open University and co-director of the OU Energy and Environment Research Unit. He has developed courses and research programmes on technological innovation, focusing in particular on renewable energy technology policy.

**David Milborrow** is an independent energy consultant, who has written widely on wind power economics and intermittency issues.

**Jonathan Scurlock** is Chief Adviser, Non-food Crops, National Farmers Union, and also a Visiting Research Fellow in the Energy and Environment Research Unit at the Open University. He writes here in a personal capacity.

**Dr. Susan Roaf** is an energy consultant and Oxford City Councillor. She has worked on solar houses, including her internationally known EcoHouse. She is a visiting Professor at the Open University and at Arizona State University.

**Dr. Rajat Gupta** is a Senior Lecturer in the Department of Architecture at Oxford Brookes University. He has been part of the Oxford Solar Initiative project, looking at the role of PV Solar. He is a Visiting Scholar in the Herberger Centre for Design Excellence at Arizona State University, USA.

**Dr. Bob Everett** is a Lecturer in the Faculty of Technology at the Open University and a member of the OU Energy and Environment Research Unit. He has wide ranging interests in the efficient conversion and use of energy.

**Dr. Horace Herring** is a Visiting Research Fellow in the Energy and Environment Research Unit at the Open University, working on sustainable energy policy issues, energy economics and environmental history.

**Dr. David Toke** is Senior Lecturer in Environmental Policy at Birmingham University. He has written widely on energy policy and carried out research on planning issues associated with wind power.

**Dr. Stephen Peake** is a Lecturer in the Department of Design and Innovation at the Open University, where he has been looking at carbon politics and policies. He is a member of the OU Energy and Environment Research Unit.

**Godfrey Boyle** is Senior Lecturer in the Department of Design and Innovation at the Open University and Director of the OU Energy and Environment Research Unit. His special interests include energy modelling and energy scenarios.

**Peter Harper** is Head of Research and Innovation at the Centre for Alternative Technology in Wales, Visiting Professor at Ritsumeikan University in Kyoto, Japan, and a Visiting Research Fellow in the Energy and Environment Research Unit at the Open University. He has written widely on environmental policy issues.

# Introduction: Sustainable Energy: The Options

*David Elliott*

Until recently, one of the main issues that has faced humanity in terms of its use of energy resources has been the worry that the majority of the resources currently relied on are finite and will inevitably become depleted by our ever-growing use of them. Oil and gas reserves are the most obvious example, but coal reserves are also not infinite. There have of course been serious *local* pollution issues of various kinds associated with use of these energy sources, but it is only in the last couple of decades that it has become clear that, due to the *global* impact on the climate system of burning fossil fuels, it may not be wise to bum off the fossil fuel reserves that remain. There are also concerns about the use of fissile fuels – not only is the uranium resource finite, but using it can involve both short-term and very long-term impacts on society and the ecosystem.

This new concern over the wider global issues is reflected in the use of the term 'sustainable energy'. The implication is that the long-term *availability* of energy sources is not the only issue. What also matters is that they can be used into the future without damaging the planets climate and ecosystem, or indeed the human social system, i.e. their use must be socially and environmentally sustainable. This social and environmental extension of the concept of sustainability was the interpretation introduced in the classic 1987 Brundtland report definition of sustainable development, which was seen as 'development that meets the needs of the present generation without compromising the ability of future generations to meet their own needs'.

'Sustainability' could be treated as an absolute requirement, meaning, in the extreme, that the only energy resources that could be considered as sustainable were those that would be available forever – for the indefinite future – *and* had zero or very low impacts. However, it is more usual to think of sustainability as being a matter of degree – so that sustainable energy resources are those that have reasonably long lifetimes and relatively low impacts. By contrast, the term 'renewable energy' is, or ought to be, more of an absolute.

Strictly the term should be reserved for those energy sources which really are likely to be available for the indefinite future – energy sources such as the winds, waves and tides, which are naturally regenerated. These energy sources will continue to be available as long as the planet receives solar energy input, and as long as the moon remains in orbit. By contrast, although it has been claimed that some hydrocarbons are still continually being produced deep underground, it is generally agreed that the bulk of earths' fossil fuel inheritance was created at specific points in the earths geological history. In addition, the planets' reserves of fissile material are also not being renewed – they were bequeathed at the planets birth.

These definitions of terms may seem rather unimportant, except for the fact that, for example, in both the US and UK, there have been attempts to label nuclear energy as a renewable resource. In addition, the terms 'renewable' and 'sustainable' are also often used interchangeably, and, along with the even looser term 'green energy', these terms are sometimes used simply to imply that the source has good environmental credentials. This can be confusing. For example, although it is true that, in most cases, the use of renewable energy sources only involves relatively low environmental impacts, that is not automatic – some may have significant local impacts. Similarly, although in normal operation, nuclear plants may have relatively low environmental impacts, few environmentalists would see them as green energy technologies, and certainly uranium is not a renewable resource.

This book attempts to avoid confusion by adopting the tight definition outlined above for renewables (as technologies reliant on indefinitely available energy resources) and the wider and more flexible definition for sustainable energy (sources with relatively long life and relatively low impacts). On this basis, nuclear power might arguably be sustainable for some period ahead, but it is not renewable. Similarly, fossil fuel use with carbon capture and storage might be sustainable for some period ahead, but this approach does not make fossil fuel renewable.

The term 'green energy' is in effect a catch all, often used to cover energy from any technology that has some environmental benefit. Moreover, 'green energy technologies' can be relevant to both the supply side and the demand side – since reducing demand or using energy more efficiently can reduce environmental impacts. Once again, as with the term sustainable, there will be degrees of 'green-ness'.

## The technological options

Although the criteria for being a fully 'green' energy source (long-term sustainability and fully renewable availability) may seem onerous, in fact there is no shortage of candidate technical options for reducing carbon emissions – on the supply side and the demand side. This book looks as some examples of both.

In order to review the field, this book draws together a range of papers from leading authors looking first at the key sustainable energy options and then at how their use might be spread. It focuses on the UK and EU, but its analysis is relevant globally. Indeed, in summarising current thinking in the UK and EU, it should be of interest to researchers, activists and decision makers around the world – including in the US and the newly industrialising Asian countries. The emphasis is both on the problems as well as the opportunities. While utopian visions can be useful to motivate change, there is also a need for hard headed assessments of what can be done in practice and how problems can be overcome. At the same time there is a need to discard some of the myths and misconceptions that have grown up about what is possible and what is not.

Perhaps one of the main miss-readings is the common confusion of 'energy' and 'electricity' – these words are often used interchangeably in the media and even in public debates. Only around a third of the energy used in modern economies is supplied as electricity – most of the rest is used as heat and as transport fuel. To some extent, this confusion is due to a historic focus in the west on technologies which supply electricity, often very inefficiently. This book attempts to avoid that bias by looking at examples of heat supply and use, including solar heat. Various approaches to the efficient use of energy are also discussed, not least since changing patterns of energy use and new approaches to demand management will have major implications for the supply side.

The emphasis is nevertheless mainly on the supply side, in part since dealing fully with all the many issues and options on the demand side would require at least another book – for example on demand management and transport policy issues. Even within just the supply side, a degree of selection has proved to be necessary. For example, this book does not look in any detail at hydro electric power, despite this being a very large renewable energy source, supplying around 18% of world electricity. In part, this absence might be justified by the argument that large hydro is very well established, whereas this book focuses on the newly emerging renewables. In addition, this book focuses mainly on the UK, where there is little potential for significant expansion of hydro. Nevertheless, it is important to realise that hydro, particularly micro hydro, has a very important role to play in the developing world. There is also around 9 giga watts of geothermal electricity generating capacity in place globally and in some locations geothermal aquifers are used as heat sources. However, this option is not currently being taken up in the UK, and is not covered in this book, apart from mention of the use of ground source heat pumps for heating buildings, and strictly they do not use geothermal heat, but rather solar derived ambient energy pumped into a building by the use of electricity or gas. It is also true that geothermal energy is not strictly a continuously renewable source – the local heat gradient is exhausted after a period, although it would be renewed over time. Biomass is however covered, focusing on biofuels for transport, although strictly this too is not continually renewable – energy crops are grown and harvested in batches. In terms of the supply side though, the main emphasis in this book is on wind, wave, tidal and solar power – the so called 'natural flow' renewables.

By looking at the prospects for these new sources, this book attempts to review the potential and problems of moving towards a long-term fully sustainable energy future, based initially on the adoption of more efficient generation and use of fossil fuels, and the sequestration of carbon emissions, in parallel with the expanding use of renewables, as a preliminary to a full transition to renewables. The emphasis is on the later phase, i.e. mainly on the newly emerging renewable energy technologies. For that reason, technologies such as Combined Heat and Power and Carbon Capture and Storage, which may make the use of fossil fuel more sustainable in the short to medium term, are not covered in any detail. Similarly, nuclear power is not discussed in detail, apart from in the opening chapter, where it is argued that its role as a transitional technology is as problematic as its potential as a long-term energy source. These issues were explored in detail in another book in this series – Nuclear or Not? The wider and longer-term implications of the use of hydrogen as a new fuel and energy transmission vector is also touched on in passing, and in the conclusion, but is not explored in any detail. That deserves a book on its own.

The practical development of the necessary technologies is of course inevitably only the start – the real issue is how to successfully deploy them into wide-scale use. A range of support strategies have emerged around the world aimed at speeding up full-scale deployment and diffusion, some of them based on providing interim subsidies, others on making use of carbon trading arrangements to simulate uptake of low carbon options. While a full assessment of the various approaches to supporting renewables and other sustainable energy options would require yet another book, some examples are reviewed critically in this study, to indicate some of the problems and hopefully some of the solutions.

The various technological options looked at in this book are of course only part of the solution to climate change. Given major commitments to energy efficiency and the increasing use of renewable energy sources, it should be possible to reduce emissions significantly. However, although the renewable energy sources will last indefinitely, there are limits to how much energy can be extracted. These limits may be some way off, but if demand for energy continues to grow they will become real. Clearly growth in energy use cannot continue for ever on a planet with a finite solar energy input. Unless we look elsewhere for new sources, there will ultimately be a need to limit the growth of human material expectations.

This implies a longer-term process. However, some environmentalists say that, far from this being a long-term issue, we have already reached the point when the impacts of economic growth, and consequent energy use, are undermining the planets' ecosystem. Moreover they claim that, limiting demand now would make it easier to meet our needs from renewable energy sources, and would make the transition to a fully sustainable future easier. We therefore need, they say, to develop a sustainable approach to consumption, and that, if the climate change problem turns out to be as serious as some now predict, then this type of social and cultural adjustment may actually be urgent. This issue is taken up in the penultimate chapter, which looks at what lifestyle changes might deliver in terms of reduced environmental impact.

As noted earlier, the emphasis in this book is mainly on experiences in the UK and EU. However the introductory chapter on nuclear and renewables includes a review of the wider global context, Chapter 11 looks at a range of global energy scenarios, while the final chapter takes a brief look at the wider global implications for making a transition to a sustainable energy future. Hopefully the global issues, and many of the other wider policy issues touched on in this book, will be developed more thoroughly in subsequent books in this series. Indeed, rather than being an even partial 'blueprint', this book might be seen as simply laying out some of the technological options to be considered in the rapidly expanding area of policy debate on how to move towards a sustainable energy future globally.

## Note on units

**Power units:** The power using or generating capacity of devices is measured in watts, or more usually kilowatts or kW (1 kW = 1,000 watts). Larger units are megawatts or MW (1,000 kW), giga watts or GW (1,000 MW) and tera watts or TW (1,000 GW).

**Energy units:** the kilowatt-hour (kWh) is the standard unit by which electricity is sold – 1 kWh is the energy produced/consumed when a 1 kW rated generator/energy consuming device runs for one hour. A megawatt-hour (MWh) is 1,000 kWh. Similarly 1,000 MWh = 1 GWh and so on.

However, some analysts sometimes use the basic physical unit for 'work', the **joule ('J')** or multiples of joules. One watt is one joule per second, so a kWh is 3,600,000 Joules, and the joule is thus a very small unit. Hence large multiples are common, e.g. peta joules or 'PJ' (1,000 Tera joules) and exa joules or 'EJ' (1,000 Peta joules).

# Part I

# The Sustainable Energy Options

# 1
# Sustainable Energy: Nuclear Power and Renewables

*David Elliott*

## Introduction

If we are to move towards a sustainable approach to energy generation, that is, an approach which can be continued into the future without undermining the planetary ecosystem, then greenhouse gas emissions must be reduced or, ideally, eliminated. However that is not the only criterion for environmental sustainability. Other emissions must also be considered, including the emission of acid gases and the release of radioactive materials. In addition, for some biologically-based energy technologies, there are concerns about the impacts on biodiversity, and indeed (e.g. if large-scale biomass developments are allowed) on ecological viability.

The relative importance of these various emissions and impacts is a matter of some debate. For example, some argue that long-lived radiation emitters present unique hazards to biological systems so that significant releases must be avoided at all costs. Others however feel that the social and economic damage that could be caused by climate change will far outweigh any conceivable damage caused by the use of nuclear power. This view is sometimes used to support the claim that we should resuscitate nuclear power as a major energy option.

This chapter attempts to explore that contention by comparing the relative prospects for nuclear and renewables, and how they might interact, focusing in the main on the UK context, but widening out in the concluding section to include the global context. The analysis of the UK energy options assumes a major commitment to energy efficiency, including the development of Combined Heat and Power using fossil fuels more efficiently, as a common and crucial feature for any sustainable energy future. So demand/efficiency issues are not explored here – they are in any case the topic of other chapters. Instead the focus is on the non-fossil supply options – nuclear and renewables.

## Is there enough energy for the future?

The first point to check is whether they both actually are sustainable options, in the sense of being able to provide energy for the long-term future. For the natural-flow renewables, that is basically no problem – solar heat and light, the winds, waves and tides and river flows will continue more or less unchanged as long as this planet survives. Climate change and other human planetary modifications may lead to slight variations in the availability of some of the climate and weather related flows. There are also limits to how much energy we can extract from these flows, and from tidal flows, without having significant impacts of the ecosystem, especially in local terms. However, overall, these limits should still allow us, in time, to be able to extract sufficient energy to meet global energy demands of at least twice the current level and possibly much more. Indeed, some argue that renewables, including biomass and PV solar, could ultimately meet 10 or even 20 times current global energy demand – on an indefinite basis (Elliott, 2005).

The technical and economic viability of using the various renewable energy sources is of course a matter for debate. Some are currently expensive, but it is claimed that new technologies can make renewables more competitive, and some are already competitive. There is also the issue of intermittency, although as explored in Chapter 2, in operational practice, assuming a diverse range of renewables feeding into grid networks, this just translates into an extra cost, for the provision of backup or storage facilities (Milborrow, 2001). But even assuming significant economic constraints, and also land use and operational constraints, the renewable resource is clearly large and, in effect, available indefinitely into the future.

By contrast, the long-term prospects for nuclear fission using uranium are less clear. They depend on an inevitably finite uranium resource, possibly augmented at some point by the use of thorium. Within that larger constraint, resource estimates are dependent on economic considerations – for example, at present prices, known uranium reserves are considered sufficient for only a few decades ahead, but if prices rise then further reserves are likely to become available. Longer term, ores with lower uranium concentrations may become viable, and if energy costs rise very dramatically, it may even be possible to use the very low concentrations of uranium that exist in sea water. On that basis, it has been suggested that the resource could be relied on for many centuries and possibly even longer. However, for the foreseeable future, the nuclear option must rely on uranium ore, and reasonably economically viable reserves of that are much more limited. Indeed, at various times, the nuclear industry has used concerns about the long-term availability of fissile material to justify the development of Fast Breeder technology, to improve the efficiency of use of uranium. In 1990 the UK Atomic Energy Authority pointed out in a booklet on the greenhouse effect, that a nuclear programme providing up to around 50% of world electricity requirements by 2020, would result in a 30% reduction in global carbon dioxide emissions from what they would otherwise have been, but warned that if a programme of this sort was based just on conventional 'burner' reactors, 'the worlds uranium supplies that are recoverable at a reasonable cost would be unlikely to last more than about fifty years' (Donaldson *et al.*, 1990).

Subsequently, with the closure of most Fast Breeder projects around the world, and more uranium finds, less is heard of this argument, although Fast Breeders are sometimes included in the so called 'Generation 4' range of possible new technologies for nuclear generation. The claim is that Breeders could help stretch uranium reserves so that, depending on the rate of use, they would last for many hundreds of years, possibly a thousand years or maybe even more. It has to be said however that this would require a very large programme of reactors and reprocessing plants, and at present the costs are unknown, and there would clearly be major plutonium proliferation and waste disposal implications (Elliott, 2003a).

To summarise, in terms of the basic resource, in theory the fission option seems to have a time horizon of the order of 'hundreds of years', depending on rates of use, the level of fuel costs that were acceptable, and the technology used – the breeder reactor could stretch it. Even so, uranium-based nuclear fission is not basically a sustainable option for the long term, and there may even be medium-term limits – if a very large programme was launched, seeking to make a significant contribution as a response to climate change, then the resource constraints could become significant (Mobbs, 2005).

## Is the energy carbon free?

The second key question is, can renewables and nuclear actually deliver carbon-free power? The natural flow renewables certainly do not generate any greenhouse gases in their operation (they do not use any fuel), and biomass use can be considered to be more or less greenhouse gas neutral if the rate of use is matched by the rate of replanting. However, in both these cases, building the energy conversion technologies and fabricating the materials for them will involve energy use, and in the case of biomass, energy is also used in harvesting and transporting the fuel.

Since most of this energy will for the moment come from fossil-fuelled power plants, there will be associated greenhouse gas generation. The scale of these associated emissions will vary depending on the technology, with each having different input energy requirements/kWh generated. Table 1.1 gives some basic emission data for various renewables. Table 1.2 presents another set of data including estimates of emissions from nuclear, coal and gas plants. As can be

***Table 1.1*** **Life-cycle emissions from renewables g/kWh**

| | Energy Crops | Hydro Small-scale | PV solar | Solar Thermal | Wind |
|---|---|---|---|---|---|
| $CO_2$ | 30–40 | 8.6 | 98–167 | 20–30 | 6.5–9.1 |
| $SO_2$ | 0.08–0.16 | 0.03 | 0.20–0.34 | 0.50 | 0.02–0.09 |
| $No_x$ | 1.1–2.6 | 0.07 | 0.18–0.30 | 0.23 | 0.02–0.36 |

*Source*: B. Norton, P. Eames, S. Lo, *Renewable Energy* 15 (1998) pp. 131–6.

***Table 1.2*** **Emissions produced by 1 kWh of electricity based on life-cycle analysis**

| | Greenhouse gas gram equiv. $CO_2$/kWh | $SO_2$ milligram/ kWh | $NO_x$ milligram/ kWh |
|---|---|---|---|
| Hydro | 2–48 | 5–60 | 3–42 |
| Coal (modern) | 790–1182 | 700–32321 | 700–5273 |
| Nuclear | 2–59 | 3–50 | 2–100 |
| Natural gas (CCGT) | 389–511 | 4–15000* | 13–1500 |
| Biomass (forestry waste combustion) | 15–101 | 2–140 | 701–1950 |
| Wind | 7–124 | 21–87 | 14–50 |
| Solar PV | 13–731 | 24–490 | 16–340 |

* top of the range is for rarely used high-sulphur 'sour gas'.

*Source*: *Hydropower-Internalised Costs and Externalised Benefits*; Frans H. Koch; International Energy Agency (IEA) – Implementing Agreement for Hydropower Technologies and Programmes; Ottawa, Canada, 2000.

seen there are some disagreements between these data, notably the much higher figures quoted at the top end of the range for wind and solar, which is only partly explained by the fact that the data in the first column of Table 1.2 is for *all* greenhouse gas emissions, converted to carbon dioxide equivalent.

Disagreements like this, over the relative greenhouse gas emissions from nuclear and renewables, bedevil this field. For example, another study quotes greenhouse gas emission for nuclear in the range 9–100 grams of $CO_2$/kWh and for wind 11–75 g/kWh (Holdren and Smith, 2000) and another study has even claimed that the emissions/kWh of net energy from nuclear plants could be larger than those from high efficiency natural gas-fired co-generation plants (Fritsche and Lim, 2006).

Some of these differences evidently relate to different accounting assumption e.g. concerning the operational lifetimes of the plants, and the mix of fuels used to generate the electricity used for producing the materials used in the construction of the plants and for the production of the fuel for the nuclear plants. However it is generally agreed that, under present conditions, both nuclear and renewable plants have significantly less associated emissions than fossil fuel-based plants. The Intergovernmental Panel of Climate Change put it like this: 'The life cycle GHG emissions per kWh from nuclear power plants are two orders of magnitude lower than fossil-fuelled electricity generation and comparable to most renewables' (IPCC, 2001).

For most flow renewables, the total input energy requirements/kWh generated are certainly low, the main exception being photovoltaic solar, which has the highest energy input/output ratio. That also translates into capital and generation costs, with PV coming out highest. That is not surprising since the PV fabrication costs (including the energy costs of fabrication) are the main capital costs, and capital costs are the main element of generation costs – there are no fuel cost.

In the case of nuclear plants, the associated emissions and the relatively high generation costs, are not simply linked to plant construction. In addition, there is the energy needed for uranium mining and processing the fuel e.g. to produce enriched uranium. The latter is a very energy intensive and expensive activity – the cost of the fully fabricated fuel represents around 20% of the final electricity generation cost. The energy efficiency of the separation/concentration process can be improved with more advanced technology: centrifuge methods are much less energy intensive than the diffusion processes so far mostly used. But it is hard to see how improvements in fabrication efficiency could continually compensate when lower and lower quality ores have to be used. The high grade ores currently used contain around 2% of uranium (20,000 parts per million), the lower grade ores only 0.1% (1000 ppm) while granite contains just 4 ppm. Although the energy requirements will not increase in direct proportion, if lower grade ores have to be used, the net energy requirement could be a significant proportion of the energy the plants produce. Indeed, it has been argued that, at some point, if a very large nuclear programme was launched in response to climate change, the energy needed for the exploitation of leaner ores could require more input energy from fossil fuels than the nuclear powerplants will provide. As a result, the net associated emissions would be larger than those produced if the fossil fuel was simply used to generate electricity for general use, rather than for refining/fabricating increasingly lower grade uranium ores (van Leeuwen and Smith, 2005).

The assessment of when the so-called 'point of futility' is reached, when more energy is needed to mine and process the fuel than is produced by the reactor, depends on a variety of complex factors, including the energy efficiency of the fuel fabrication and enrichment processes, and how this energy is provided. Certainly the World Nuclear Association has claimed that Leeuwen and Smiths' analysis is faulty, and has estimated that, even with lower grade ores, the input energy required over the full plant lifecycle is less than 3% of the total energy produced (WNA, 2005).

There are clearly some disagreements on this issue, some of them statistical and methodological, a situation which led the UK House of Commons Environmental Audit Committee to suggest a review by the Royal Commission on Environmental Pollution (EAC, 2006). For example, another study of energy outputs to energy inputs over the complete life cycle, indicated that, at present, nuclear plants only generate up to 16 times as much energy as is required to build them and produce their fuel (Gagnon *et al.*, 2002). For comparison, this study indicated that wind turbines can produce up to 80 times as much energy as is needed for their construction – in part since there are no direct fuel requirements. Moreover, this figure is likely to be improved as new technologies emerge, while as we have seen, some analysts claim that, even with new technology, the figure for nuclear is likely to fall if lower grade ores have to be used.

Of course, if non-fossil energy sources are available, then this energy/carbon debt problem would be avoided (SDC, 2006). On this basis some have suggested that, in the longer term, renewable sources could be used to produce carbon-free uranium fuel, for example using wave energy to separate it out from sea water, despite the very low uranium concentrations (0.0003 ppm). Equally, in theory, nuclear sources could be used to provide this energy, in which case the nuclear cycle would be self-sustaining, for as long as the uranium resource was available.

A parallel option might be to make use of another resource – thorium. Little interest has been shown in this material so far, but there are relatively large reserves, although using it as a nuclear fuel presents some problems since it is not directly fissile – it would need to be used in conjunction with plutonium in a breeder-type reactor (Price and Blaise, 2002).

In the longer term, it may be possible to switch over to nuclear fusion, for which the fuel is likely to be far less constrained, although, depending on the technology used and the scale of operation, there could still be significant resource limits e.g. lithium reserves (used for the production of tritium) are not large, at least not on this planet. Even so, it is usually argued that fusion could produce power for many centuries ahead, although at unspecified economic costs and with unknown total life cycle energy/carbon costs. It is certainly likely to be a very capital-intensive option, with the energy costs likely to be high. Given that, on the basis of current designs, fusion reactors would contain a radioactive plasma at 200 million degrees Celsius, there are also safety risks with this technology. In addition, the high neutron fluxes involved would induce radioactivity in the materials in the reactor, which would have to be stripped out and stored. So there would still be active wastes to deal with, although with much shorter half-lives than fission products (Elliott, 2003b).

To summarise, renewables could provide significant amounts of low carbon energy on an indefinite basis, while nuclear fission could produce energy with possibly a slightly higher carbon content than most renewables for a period depending on the rate and scale of use – if fossil fuels were used to produce the reactor fuel, emissions would rise as lower grade uranium had to be used, and the overall this option is resource limited. The prospects for the fusion option are unclear, but it is possible that it could provide low-carbon power on a longer-term basis, although with some operational problems and at unknown cost.

## Heath, safety and environmental impacts

As noted earlier, reduced carbon/greenhouse emissions are not the only issue. The nuclear fuel cycle inevitably involves the risk of release of radioactive materials, most obviously following major accidents, with the possibility of a large number of deaths. Of course, hydro electric dam failures can also lead to major loss of life. Indeed some comparisons suggest that the overall mortality/GWh figures for nuclear are much less than for hydro. See Table 1.3 for data from the Australian Uranium Information Centre.

However the use of 'immediate fatalities' as a measure in this comparison rather limits the significance of the assessment. The 31 deaths shown are presumably these amongst staff at Chernobyl in 1986, whereas subsequently many thousands of early deaths amongst clean up workers and the public have been attributed to this accident. The exact number is still the subject of some uncertainty and debate, in part due to the fact that it takes time for cancers to develop. A '10 year after' review suggested that around 2,500 of the 200,000 'liquidators' who were brought in to clean up the plant, might develop cancers, while a further 2,500 of people from the immediate area might also develop cancers (IAEA, 1996). A follow-up UN assessment in 2000 identified 1,800 cases of thyroid cancer in children, although these were seen as treatable e.g. by surgical removal, and it was claimed that, overall, there was 'no evidence of a major public health impact' (UNSCEAR, 2000). A '20 years after' study, produced by the Chernobyl Forum, involving eight UN agencies including the WHO, and the governments of Russia, Belarus and Ukraine, suggested that a total of 4,000 deaths will probably ultimately be attributable to the accident, but that there has been no observed long-term rise in the incidence of leukemia, or any detectable decrease in fertility or increase in birth defects (Chernobyl Forum, 2005).

***Table 1.3*** **Comparison of accident statistics in primary energy production 1970–92**

| Fuel | Immediate fatalities | Who? | Deaths per GWy* |
|---|---|---|---|
| Coal | 6400 | workers | 0.32 |
| Natural gas | 1200 | workers & public | 0.09 |
| Hydro | 4000 | public | 0.80 |
| Nuclear | 31 | workers | 0.01 |

*Basis: per 1,000 MWe operating for one year, not including plant construction, based on historic data which is unlikely to represent current safety levels in any of the industries concerned.

*Source*: http://www.uic.com.au/ne6.htm#6.3

These studies focused on the most heavily contaminated areas, whereas, although the contamination was less, other areas with large populations were also impacted. A subsequent study by the WHO-based International Agency for Research on Cancer (IARC, 2006) suggested that the eventual figure across Europe as a whole might be around 16,000 attributable cancer deaths, but with an uncertainty range of 6,700 to 38,000, while an independent study has claimed that the global range could be 30,000–60,0000 deaths (TORCH, 2006).

Certainly reports of major problems have continued. The Children of Chernobyl charity has noted that the Clinical Institute of Radiation Medicine in Minsk claimed that the cancer rate has risen in Belarus by 40% between 1990 and 2000 and in the highly contaminated Gomel region, by 55%. Moreover, as Dr Rosalie Bertell has noted, in any case, it is not just deaths, but also major illnesses that worry people (WISE Nuclear Monitor 634). However, some UN studies have claimed that some of the health effects that had been attributed to the accident might have had psychosomatic causes or be due to the stress resulting from over-zealous relocation of people out of the contaminated area (UNDP/UNICEF, 2002). The 2005 UN/Chernobyl Forum report suggested that 'the mental health impact' was 'the largest public health problem unleashed by the accident'.

Clearly there are still disputes about the statistics, not made any easier by the break-up of the USSR shortly after the disaster. There are also differing interpretations of what represents a health problem and different perceptions of risk. There have so far been a small number of major nuclear accidents and, statistically, the associated risks of serious impacts, averaged out across the population, are relatively low. However, in this analysis, we have not yet included any estimates of risks due to intentional acts of sabotage or terrorism. A successful attack on a major nuclear facility, for example by crashing a fully-fuelled civil airliner into a reprocessing plant where large amounts of highly active materials are stored, could lead to the release of significant amounts of radioactive materials and cause many casualties (STOA, 2001). Equally, there is the risk of illegal diversion of material that could be used for making weapons, and more generally the spread of nuclear technology has major implication for nuclear weapons proliferation. Whether that can or should be factored in to the nuclear risk assessment is a matter of debate. But certainly, by contrast, if we are to make fair comparisons, renewable energy projects (large hydro possibly excepted) are unlikely to be the target for terrorist attack and offer no opportunity for weapons production.

Even leaving major impacts due to accidents, attacks or theft aside, the debate over the heath and safety implications of the various energy technologies is a complex one. There will be occupational hazards with renewables as with any technology: so far there have been around 20 deaths amongst installation/maintenance workers in the wind industry world wide. There are also debates about the long-term implications of low level radiation releases from nuclear facilities and the impacts of long-term waste repositories. In a recent full life cycle analysis of energy systems the World Energy Council warned that, while in terms of greenhouse gas emissions 'renewable fuels and sources and nuclear compare favourably ...some of the externalities cannot be covered by the LCA methodology' – for example long-term impacts from nuclear waste releases. These it says 'must be addressed within the political process' (WEC, 2004).

This point has particular relevance for renewables, given that the significance of some of the impacts is often a matter of subjective assessments, for example, with regard to disruption of scenic views. Some of the other impacts may be more amenable to objective assessment, for example the local social and environmental impacts of large hydroelectric dams and the impacts on biodiversity of large biomass plantations. Even so, it is hard to put these various impacts in a common framework and to set them against the social and environmental impacts that may occur as a result of climate change. Attempts have been made to use cost-benefit analysis, based on putting cash costs on the impacts. In one such approach, the EU's EXTERNE programme, 'cost adders' were produced for the extra cost that should be added to the generation costs to reflect the full life-cycle social and environmental costs, including mortality and injury costs. See Table 1.4.

***Table 1.4*** **Extra cost (averaged figures) resulting from social and environmental damage in euros/kWh**
to be added to electricity costs – assumed as 0.04 euro/kWh average across the EU in Euros/kWh

| | |
|---|---|
| Coal | 0.057 |
| Peat | 0.035 |
| Gas | 0.016 |
| Biomass | 0.016 |
| PV solar | 0.006 |
| Hydro | 0.004 |
| Nuclear | 0.004 |
| Wind | 0.001 |

*Source*: EXTERNE: Externalities of Fuel Cycles. Externe project DGXII, EC, 1998–2001.

As can be seen, wind comes out as having four times less impact than nuclear. However these average figures have to be treated with caution – at best they are only indicative. In particular, the EXTERNE team did not include estimates for the full long-term cost of climate change in their analysis, since this was seen as impossible to calculate with any accuracy. Moreover, quite apart from the difficulty of putting costs to the continuing, long-term, social and environmental costs of climate change, it is hard to put prices on intangibles like the relative value of a life compared with a view.

The 2006 Stern report on the *Economics of Climate Change*, produced for the UK Treasury, did try to address some of these issues, for example by adopting an approach to discounting future costs and risks that was more sensitive to the interests of future generations than is usual in conventional short-term economics (Stern, 2007). However it was a very general approach looking at global trends and costs. It did suggest that the (damage) costs of climate change would be much higher than the (control) costs of taking action to mitigate emissions (by perhaps a factor of 20), but it did not look in any detail at the relative emissions of the various mitigation options.

There have however been some smaller, more focused studies, which, despite the methodological problems, nevertheless provide some suggestive insights. For example, a detailed EXTERNE-based study by AEAT of the externalities associated with the UK's power plants, estimated that, in terms of global warming related emissions, the mid-range damage cost for nuclear were around 0.37 mECU/kWh, while those for wind were 0.25 mECU/kWh (AEAT, 1998). That implies that nuclear has around 50% more impact than wind. However, as AEAT pointed out, there were many uncertainties in the estimates. In general, although useful as a way to identify overall patterns, EXTERNE type studies are often seen as having limitations (Krewitt, 2002).

A more general approach to assessing relative impacts is to consider wider ecological principles, such as the idea that we should avoid introducing major irreversible changes. For example, taking a long-term perspective, it could be argued that the release of radioactive materials, some with active lifetimes of several thousand years, represents a serious and irreversible burden to the ecosystem, which should be avoided at all costs as a matter of principle. By contrast, most renewable energy technologies have relatively small local environmental impacts and, if necessary, the equipment can be removed leaving at most only very minor short-term damage.

However, general assessments like this can often be countered by specific technical arguments. For example, it can be argued that the volumes of nuclear waste involved are not large and that engineered containment can reduce the risks significantly. Moreover, it is hard to escape the wider strategic issues. For example, while it may be admitted that, even with the best possible technology, there will inevitably always be a risk of accidental releases from the nuclear cycle, it is also sometimes claimed that the resultant impacts on life and health will be small relative to those from climate change.

This may be a little disingenuous. After all, it could equally well be argued that if, instead of funding a nuclear expansion, we funded renewables seriously, we could obtain the same benefits in climate change terms, without the long-term problems of radioactive wastes. Moreover, if the renewables with large potential impacts were avoided, we could possibly get a significantly better overall outcome in social and environmental terms.

## Economic comparisons

General arguments like this may go some way to helping us identify views on the relative potential merits of each option in abstract, but in practice decisions are usually made on the basis of strategic political and economic concerns, with direct cost still often being the defining factor. Certainly, in a fully competitive market-defined context, the cheapest options would be likely to dominate. However, although liberalised, the UK energy market has been modified and adjusted to support some options on strategic grounds via a range of market enablement mechanisms and the provision of subsidies for desired options. At present, these include the provision of grants for new renewable energy technology development and deployment, the Renewables Obligation cross subsidy, and loans to compensate for the economic problems experienced by the nuclear industry.

Given the impacts of past and current subsidies, it is often hard to produce reliable figures for the costs of the various options. The debate over nuclear power has certainly been bedevilled by conflicting economic interpretations of historic and predicted costs. For an interesting if lengthy review, see the House of Commons Energy Committee's Fourth Report, Session 1989–90 on 'The Cost of Nuclear Power' (June 1990). This includes estimates by Lord Marshall, who was chairman of the Central Electricity Generating Board until its demise, of an average price of electricity from the UK's Advanced Gas Cooled Reactors of

7.5 p/kWh, and 6.25 p for the US Pressurised Water Reactor design. The Select Committee put the latter at more like 7.34 p/kWh. More recently however the nuclear industry has argued that new technology, such as the AP 1000 reactor concept, could generate at around 2.5 p/kWh, a claim that seems to fly in the face of UK experience that reactor costs have continually risen. Of course more radical designs may emerge, like the Pebble Bed Modular Reactor concept, but so far that is speculative.

One way forward is to use a parametric approach, based on 'learning curves' of generation cost reductions against delivered capacity, to look to the likely future for the various energy options. On this basis, a study by the Performance and Innovation Unit carried out for the UK Cabinet Office in 2002, found that the long-term prospects for most renewables looked better than for nuclear: see Table 1.5. (The results of the PIU's learning curve analysis are discussed in more detail in Chapter 9).

Clearly there is a lot more to say on the issue of costs and subsidies, and these issues are explored in some subsequent chapters, but to some extent the question of price reflects decisions about the level of support given to the various options. It may be a sad comment on economics, but, given that a level playing field does not exist, the price of any option will depend to some extent on political decisions.

One thing however is clear, whatever happens there will be a need to replace all the existing generating plant over the next few decades as it reaches the end of its operational life. A study in 2006 by Deloitte consultants suggested that the cost of moving to a carbon-free UK energy system would be around £50 billion. They suggested that a replacement programme for nuclear power stations that come to the end of

***Table 1.5*** **Cost of electricity in the UK in 2020 in pence/kWh**

| | |
|---|---|
| On Land wind | 1.5–2.5 |
| Offshore wind | 2–3 |
| Energy crops | 2.5–4 |
| Wave and tidal power | 3–6 |
| PV Solar | 10–16 |
| Gas CCGT | 2–2.3 |
| Large CHP/cogeneration | under 2p |
| Micro CHP | 2.3–3.5 |
| Coal (IGCC) | 3–3.5 |
| Nuclear | 3–4 |

*Source*: Performance and Innovation Unit, 'The Energy Review' UK Cabinet Office, 2002. http://www.pm.gov.uk/output/Page4250.asp

their life would cost up to £20.5 billion for ten new plants, with the rest being for renewables/Combined Heat and Power/ Fossil fuel use, with Carbon Capture and Storage (Deloitte, 2006). On the basis of the PIU estimates above, it is not clear however, what the cost ratio between these various elements should be: for example, renewables could be less costly than nuclear.

Moreover, as I have argued at length elsewhere (Elliott, 2007), given inevitably limited financial and technical resources, trying to back both nuclear and renewables might be problematic. In the past nuclear has tended to absorb the bulk of the funding available and many people in the renewables community fear that this would happen again, and that renewables would be sidelined. Even if that was avoided, there is still a risk that we would do neither well.

It would arguably make more sense to make a clear decision and it could be argued further that, since nuclear has had many decades of major funding, it would be reasonable to allow renewables to show what they could do, especially since the UK has some of the worlds best renewable energy resources. For example, in 2004, the Carbon Trust/DTI Renewable Innovation review estimated that by 2050, the UK could be getting 53–67% of its electricity from renewables, if the necessary support was provided. In parallel there is the carbon capture and storage option which, if successfully developed would allow fossil fuels to make up the remainder, without adding significantly to emissions, while renewables continued to expand. On this basis, going back to nuclear power could be seen as an expensive diversion. Sequestration may also prove to be relatively expensive, but it would seem easier to contemplate storing carbon dioxide gas in depleted oil and gas wells than trying to find sites for repositories for the very long-term storage of nuclear wastes.

Of course the situation elsewhere in the world maybe different. Most of the rest of Europe has significant renewables resource, but in the newly industrialising countries of Asia it might be argued that it will be hard to support continued rapid economic expansion with renewables. However it can equally well be argued that it would be preferable for them to focus on cleaning up fossil fuel use with carbon sequestration, plus renewables, rather than going down the nuclear route. The next section looks at these wider global issues.

## Nuclear and renewables: beyond the UK

So far, this chapter has focused mostly on the situation in the UK, comparing the prospects for and views on nuclear power and renewables. This section attempts to set the discussion in the wider global context, for clearly UK energy policy issues and choices cannot be reviewed in isolation from wider trends and developments elsewhere may make UK domestic policies and programmes irrelevant. For example, rapidly developing countries like China seem bound to rely increasingly on their coal reserves, and some are also trying to expand their nuclear programmes. It could be argued that, in this context, investing in clean coal combustion combined with carbon sequestration would be better than nuclear as an interim option, while developing renewable energy resources for the longer term. It is interesting that China already generates about the same amount of total renewable energy as the US, if large hydro is included. Clearly what happens in countries like China – and India – will shape the global energy and environmental future.

The Renewable Energy 2004 conference in Bonn discussed scenarios in which renewables supplied 50% of total global *energy* (heat and power) by 2050 (Renewable Energy, 2004). That would require rapid rates of deployment, but that seems feasible given that, for example, Germany has already installed 18 GW of wind capacity, with a construction rate of between 1–2 GW p.a., rising to 3 GW in 2002/4. This rate of build is not surprising given that, once planning agreements have been obtained and the site prepared, wind turbines can be installed in a matters of days. By contrast, given the long planning process and long construction process, it is hard to see how nuclear power could be expanded quickly world wide on the necessary scale. It is true that the US managed to achieve installation rates of around 2 GW p.a. during the nuclear boom period in the 1970s, and France did even better, but repeating that world wide would be challenging and to make a meaningful impact on climate change much more would be required.

The World Nuclear Status Report 2004, commissioned by the Green-EFA Group of the European Parliament, concluded that in order to simply maintain the current number of reactors (there are currently around 440 globally) would need planning, building and starting-up about 80 units over the next ten years, which implies one every six weeks, and another 200 units over the following ten years, that is one every 18 days (Nuclear Status Report, 2004). Going further and *expanding* the nuclear contribution would obviously require even more effort. A recent MIT report on the 'Future of Nuclear Power', pointed out that, given increasing demand, to increase nuclear powers' share from its present 17% of world electricity just to 19% by 2050 would require a near-trebling of nuclear capacity – 1,000–1,500 large nuclear plants would have to be built worldwide.

Globally, at present, 25 new reactors are under construction, including the Chinese, Indian and Korean programmes, with a combined capacity of nearly 21 GW. By 2020, on current plans, both India and China hope to have 20 GW of nuclear capacity (Worldwatch, 2004). However, given the rapid growth in energy demand in these countries, this will at best double their nuclear contribution in percentage terms. For example, China currently has eight nuclear reactors in operation supplying under 2% of current demand. By 2020, assuming the national plan is fulfilled, nuclear energy would still only meet under 4% of demand. Meanwhile, around the world, nuclear plants are being closed as part of phase out programmes, as for example in Germany, Sweden, Belgium and Lithuania, or retired, as they come to the end of their useful life. So far over 34.6 GW of nuclear plant has been closed or retired, and, although, as noted above, new capacity is coming on line, the end result is that globally the net nuclear contribution has been more or less static in recent years, with total energy output actually decreasing in 2003, although nuclear generation capacity did increase slightly in 2005/6 (by 1%), following the commissioning of four new plants and the restarting of a fifth (Worldwatch, 2006).

At the same time, however, the use of renewables is expanding quite rapidly around the world, in terms of both electricity supply and heat supply – with, for example, by 2006, over 60 GW of wind and over 70 GW (th) of solar thermal already being in place globally, plus of course large contributions from conventional hydro and biomass. In addition, increasing use is being made of biofuels for transport.

The EU's programme has a target of obtaining 12% of electricity from renewables by 2010, excluding large hydro (21% with hydro), and 23% of total primary energy, including heat and biofuels. Both China and India have major renewable energy programmes. India is now fourth in the world wind power league and has major PV solar, biomass and biogas programmes. It expects to have 10 GW of renewable capacity in place by 2012. China already has over 23 GW(th) of solar thermal capacity, and major hydro and biogas programmes. By 2020, it hopes to have 30 GW of wind capacity and 30 GW of biomass plant, with 120 GW of new renewables capacity overall – meeting around 16% of its expected electricity requirements by that time. The longer term potential is very large – for example, according to one estimate, China has the potential to install up to 400 GW of wind generation capacity by 2050 (EWEA, 2006). And the overall wind resource is even larger – it has been estimated at up to 1,000 GW, if offshore sites are included, with some estimates putting the overall wind potential even higher. One modelling study suggested that it might be up to 3,000 GW (Junfeng *et al.*, 2006).

Some approaches to the development of renewables are large scale and centralised. For example, it has been suggested that use could be made of giant focused-solar tracking mirror arrays in desert areas in North Africa, to generate steam for electricity production, with the electricity being transmitted by High Voltage-Direct Current links to Europe. Some large solar thermal-electric projects already exist in California, Spain and elsewhere, but supporters of the 'Concentrated Solar Power' concept clearly want a much more ambitious macro-regional approach. They claim that transmission losses could be around 3% per 1,000 km and that by 2050 such an approach could be supplying at least 15% of the EU's electricity and possibly much more, while in parallel also providing power for local desalination plants (TREC, 2006).

Less dramatically, some renewable energy technologies are well suited to supplying electricity, using local sources, direct to users, off-grid, and with around two billion people in the world unlikely ever to have access to grid power, this, along with the provision of non-fossil heating and transport fuels, is an important area of development. A study of China has suggested that given the size of the country and its dispersed population, small to medium sized decentralised energy systems, using local energy sources, are significantly more efficient than conventional, large-scale fossil or nuclear plants, since, they avoid the cost of having to provide grid links and also avoid the energy losses that would be incurred by sending power long distances over power grids (WADE, 2004).

Similar conclusions apply to many other developing countries, particularly in Africa. In theory small nuclear plants could be used for local electricity supply and also for bulk heat supply (if people are willing to have plants in or near urban areas) and to produce hydrogen for transport, this latter idea being on the agenda in the US, which has launched a major hydrogen-for-vehicles research programme. However, for the present, nuclear plants can only supply electricity for grid systems – which represents a sector responsible for only about 30% of carbon emissions in developed economies and less elsewhere.

In addition to efforts to get nuclear expansion going again in the US, it is likely that nuclear projects will continue in Asia, given the large increases in demand for electricity in the newly emerging economies.

That has major implications for nuclear weapons proliferation. However, unless radically new technologies emerge, it is hard to see how a major nuclear programme could be mounted globally in time to have much impact on climate change. Moreover, climate change could itself make this difficult. With rising sea levels and increasing storm intensities, coastal locations could be problematic, and inland sites would require reliable access to cooling water, which may also be increasingly hard to obtain as global temperatures rise.

Overall, even leaving aside the weapons proliferation issue, the prospects for nuclear globally look mixed. The International Energy Agency 2005 World Energy Outlook suggested that on present plans, given the retirement of old plants, nuclear capacity would peak around 2015 and decline thereafter. It may of course be that, as well as the Chinese and Indian programmes, new expansion programmes will emerge, e.g. in the EU as well as the US, but equally renewables could be the technology of choice for expansion around the world.

## Conclusions: energy choices

The UK was a pioneer in the development of nuclear power. It has also been a pioneer in relation to renewables – which is perhaps not surprising since it has ample renewable energy resources. It might be argued that it has been left too late to exploit them so as to meet the urgent need to reduce emissions. However, by itself, that hardly seems like a reason to revert to nuclear power. Instead it seems to be an argument for the more rapid development of renewables. The same is true elsewhere in the world, and, as illustrated earlier, the potential is very significant.

It is true that if renewables are not developed very rapidly then, in order to try to avoid massive climate impacts while expanding energy supplies, then the expansion of nuclear may seem necessary. However, the analysis in this chapter suggests that, given the overall resource limit, nuclear fission may only be able to play a limited short- to medium-term role. Even then, it would face some significant problems including cost, security and safety. Renewables also have their problems, including the diffuse and intermittent nature of some of the sources. This book looks at some of these problems. But renewables also have some strategic operational advantages – they are flexible, modular and able to fit into a locally embedded power system.

## Comparison of nuclear and renewables

| | Nuclear | Renewables |
|---|---|---|
| Resource lifetime | Uranium reserves ~ 100 years *at current use rates* ~ 1,000 years with FBR? | Effectively infinite resource lifetime |
| Resource scale | Currently ~ 6% of world energy, ~ 17% of electricity Could perhaps be doubled? Or trebled? i.e. to 50% of world *electricity*. But lifetime of the resource would then be limited | Currently – 6% (with hydro) – 17% of electricity Projection: 50% of world *energy* by 2050 (RE2004/ Bonn Conference) |
| Eco impacts | Infrastructure impacts, cooling water impacts, risks from very long term wastes ~ 10,000 years | Local visual intrusion and land use conflicts, some local eco-impacts (especially with biomass and large hydro) |
| Safety | Major accidents – 10,000 deaths, occasional/routine emissions ~ 100's of deaths | Generally low risk, except large hydro ~ 10,000 deaths |
| Costs | High and could rise as uranium resource dwindles, but new technology could emerge | Some high, but most are moderate, and all are falling as technology develops |
| Output | Electricity only, but could be used for direct heat or hydrogen production | Diverse sources: electricity, heat, fuels |
| Reliability | Occasional shut downs | Some rely on variable sources, so need grid integration to balance outputs |
| Supply Security | Uranium deposits limited to a few locations | Widely diffused energy sources |
| Security Risks | Significant terrorist targets, plutonium proliferation threat | No significant problems except with large hydro |

The chart above summarises the basic issues identified in this chapter. Overall it would seem that renewables have the edge. It may be that for local strategic reasons, nuclear power will nevertheless find an increasing role in some parts of the world, as an interim option, particularly if new reactor technology can be developed. But on balance the prospects for the renewables as a long-term sustainable energy option look far brighter.

## References

AEAT (1998) 'Power Generation and the Environment – a UK Perspective' AEATechnology report 3776: http://externe.jrc.es/uk.pdf.

Chernobyl Forum (2005) 'Chernobyl's Legacy: Health, Environmental and Socio-Economic Impacts', multi agency report produced for the UN. See: www.iaea.org/NewsCenter/Focus/Chernobyl/pdfs/EGE_Report.pdf

Deloitte (2006) '2020 vision: The next generation', Deloitte consultants, London.

Donaldson, D., Tolland H. and Grimston, M. (1990) 'Nuclear Power and the Greenhouse Effect', UK Atomic Energy Authority, Jan.

EAC (2006) 'Keeping the Lights on', Environmental Audit Committee, Sixth report 2005–6 session, House of Commons, London.

Elliott, D. (2003a) 'The Future of Nuclear Power' in Boyle, G. *et al.* (eds) *Energy Systems and Sustainability*, Oxford University Press.

Elliott, D. (2003b) 'Energy, Society and Environment', Routledge, London.

Elliott, D. (2005) 'Energy efficiency and Renewables', *Energy and Environment* special issue on efficiency, Vol. 15, No. 6, pp. 1099–105.

Elliott, D. (2007) 'Nuclear Power and Renewables – can we have both?' in Elliott, D. (ed.) 'Nuclear or Not?', Palgrave, London.

EWEA (2006) 'Wind Force 12' Chinese version of the European Wind Energy Associations review, produced jointly with Greenpeace and the Chinese Renewable Energy Industry Association.

Fritsche, U.R. and Lim, S. (2006) 'Comparison of Greenhouse-Gas Emissions and Abatement Cost of Nuclear and Alternative Energy Options from a Life-Cycle Perspective' paper presented at the CNIC Conference on Nuclear Energy and Greenhouse-Gas Emissions, Tokyo, November 1997, updated 2006. Öko-Institut, Darmstadt. See: www.oeko.de•Öko-Institut 2 Nuclear/CO2

Gagnon, L., Belanger, C. and Uchiyama, Y. (2002) 'Life Cycle assessment of electricity generation options: the status of research in the year 2001', *Energy Policy* 30, pp. 1267–78.

Holdren, J.P. and Smith, K.R. (2000) 'Energy, the environmental and health' in Goldenburgh, J. (ed.) *World Energy Assessment: Energy and the Challenge of Sustainability* , UNDP/World Energy Council.

IAEA (1996) 'One Decade After Chernobyl' report on a Conference organised by the International Atomic Energy Agency, together with the World Health Organisation, the European Commission and others; http://www.iaea.org/worldatom/Programmes/Safety/Chernobyl

IARC (2006) 'The Cancer Burden from Chernobyl in Europe', International Agency for Research on Cancer, World Heath Organisation, Paris. http://www.iarc.fr/ENG/Press_Releases/pr168a.html

IPCC (2001) *Climate Change 2001*: volume III, 3.8.4.2 Intergovernmental Panel on Climate Change p. 240.

Junfeng, L., Jinli, S. and Lingjuan, M. (2006) 'China: Prospect for Renewable Energy Development' paper commissioned for the UK Treasury Stern Review: http://www.hm-treasury.gov.uk/independent_reviews/stern_review_economics_climate_change/stern_review_supporting_documents.cfm

Krewitt, W. (2002) 'External Costs of Energy – do the answers match the questions? Looking back at 10 years of ExternE', *Energy Policy* 30, pp. 838–48.

Milborrow, D. (2001) 'Penalties for intermittent sources of energy' PIU Working paper, Performance and Innovation Unit, UK Cabinet Office, http://www.pm.gov.uk/output/Page4250.asp

Mobbs, P. (2005) Uranium Supply and the Nuclear Option. *Oxford Energy Forum*, Issue 16, Oxford Institute for Energy Studies, May. See: www.fraw.org.uk/mobbsey/papers/oies_article.html

Nuclear Status Report (2004) Report by Mycle Schneider and Anton Antony Froggatt, Independent Consultants, commissioned by the Green-EFA Group of the European Parliament. http:// www.greens-efa.org/pdf/documents/greensefa_documents_106_en.efa.pdf

Price, R. and Blaise, J.R. (2002) 'Nuclear fuel resources: Enough to last?', NEA updates, NEA News, No. 20.2, Nuclear Energy Agency, Paris. See: www.nea.fr/html/pub/newsletter/2002/20-2-Nuclear_fuel_resources.pdf

Renewable Energy 2004 (2004) Federal German Government Conference Issues Paper for the International Conference for Renewable Energies, Bonn, June.

SDC (2006) *The role of nuclear power in a low carbon economy*. Sustainable Development Commission, London, March www.sd-commission.org.uk

Stern, N. (2007) 'The Economics of Climate Change', Review for HM Treasury by Sir Nicholas Stern, Cambridge University Press. http://www.hm-treasury.gov.uk/independent_reviews/stern_review_economics_climate_change/sternreview_index.cfm

STOA (2001) 'Possible Toxic Effects from the Nuclear Reprocessing Plants at Sellafield (UK) and Cap de La Hague (France)', WISE-Paris report for the European Parliament Scientific and Technological Options Assessment programme, Strasbourg.

TREC (2006) TRANS-CSP report, commissioned by the German Federal Ministry for the Environment, Nature Conservation and Nuclear Safety from the Trans-mediterranean Renewable Energy Co-operation project. See: www.trec-eumena.org and http://www.trec-uk.org.uk/csp.htm

TORCH (2006) 'The Other Chernobyl Report', Ian Fairlie and David Sumner, http://www.chernobylreport.org/?p=downloads

UNSCEAR (2000) Report to the General Assembly by the UN Scientific Committee on the Effects of Atomic Radiation, New York.

van Leeuwen, J.S. and Smith, P. (2005) 'Nuclear Power: the Energy Balance', http://www.stormsmith.nl/ See also Flemming, D. (2006) 'Why nuclear power cannot be a major energy source', FEASTA paper. http://www.feasta.org/documents/energy/nuclear_power.htm

WADE (2004) 'The WADE Economic Model: China, A WADE Analysis', World Alliance for Decentralized Energy, Edinburgh, Dec.

WEC (2004) 'Comparison of energy systems using Life Cycle Analysis', World Energy Council, London.

WNA (2005) 'Energy Balances and CO2 Implications' (Nov); 'Energy Analysis of Power Systems' (Aug); and 'Critique of 2001 paper by Storm van Leeuwen and Smith: "Is Nuclear Power Sustainable? and its May 2002 successor: Can Nuclear Power Provide Energy for the Future; would it solve the $CO_2$-emission problem?" with reference to a 2005 version entitled "Nuclear Power, the Energy balance"', WNA Supplement, (Aug 2002, LCA comparisons updated 2005), World Nuclear Association web site: http:www.world-nuclear.org/
Worldwatch (2004) 'Vital Signs 2003', Worldwatch Institute, Washington DC.

# 2

# Wind Power and Similar Renewable Sources –Why Variability Doesn't Matter

*David Milborrow*

## Introduction

'What happens when the wind stops blowing?' is a question that is frequently asked. Nobody asks, 'What happens when a large thermal power station, or one circuit of the Anglo-French cross-channel link (1,000 MW) trips out?' In practice, transmission links, nuclear power stations and electricity consumers are far more of a threat to the stability of electricity networks than wind energy – and will be until wind energy provides a substantial amount of electricity supplies. The output from nuclear power stations – or any other type – is not totally predictable. Neither is the demand from electricity consumers. In consequence, all power systems have strategies to deal with mismatches between supply and demand. When variable renewables are added, what matters is the additional uncertainty in predicting the balance between supply and demand. The need for additional reserves increases only slowly as wind energy is added to a system, so there are no operational difficulties and only modest extra costs.

This chapter reviews studies of wind variability from around the world. The emphasis is on data and analyses from electricity utilities, but the underlying theory behind intermittency issues is also explained. The amount of extra reserves that are needed is quantified, together with the additional costs. Apparent anomalies between national studies are examined.

One of the reasons why wind energy can be handled without difficulty is that the effects of geographical dispersion smooth the wind fluctuations. Actual data from western Denmark are used to illustrate this effect. The types of 'reserves' used by power system operators are also discussed, together with the possible role of storage – another area where there are widespread misconceptions.

Another controversial topic is the 'capacity credit' of wind plant: the extent to which they can provide 'firm power', and so displace thermal plant. The results from studies in Europe and America are therefore reviewed and compared.

Most of the intermittency studies address the impact of wind energy, but the related issues associated with tidal, wave and solar energy are briefly discussed.

## Intermittency and wind power

The issue of wind variability remains controversial, nearly 30 years since the first studies were published that suggested it was not a major issue. In 2005, when the UK Sustainable Development Commission (SDC) published a report on wind energy (SDC, 2005) they set up a discussion forum on their web site. Almost immediately the controversy surfaced, with an early correspondent arguing:

> *To assert that no back up power sources are required is totally in conflict with experience of operators – see E. oN (Germany) Wind Report 2004 – where they say that 80% back up is required from conventional power generating stations.*

A few days later another correspondent retaliated: '*Of course wind power needs back-up. But this is not unique to wind power. Conventional and nuclear plants can "trip" and the entire production of such a plant halts immediately. This means drops of 1,000 MW in the supply side can occur within a few seconds.*'

So where does the truth lie? The first point to make is that the SDC report does not claim no additional backup is required. What it says is, '*With wind supplying 10% of the electricity, estimates of the additional reserve capacity are in the range 3 to 6% of the rated capacity of the wind plant. With 20% wind, the range is approximately 4 to 8%.*'

The second correspondent hit the nail on the head. Does wind energy need backup? – yes! Do nuclear power stations need backup? – yes! Does the Anglo-French cross-channel link (two 1,000 MW circuits) need backup? – yes! Do we add up all the individual backup requirements to determine the total backup needs for the UK electricity system? – definitely not! If we did, the electricity system would be extremely inefficient and costly, throwing away the enormous advantages that come with having integrated electricity networks.

It follows that discussion of the issues surrounding the integration of variable renewable energy sources cannot be divorced from an understanding of the operation of electricity networks, so this is discussed first. Characteristics of wind energy are then examined before integration issues are analysed. Related topics, including 'capacity credits', storage, local issues and the overall extra costs of integrating wind energy into a network are then covered. The emphasis is on wind energy, as most work has focused on this technology, but brief comments on the other technologies are included.

## Electricity systems

The way that electricity systems operate to meet consumer demands is similar, the world over. Consumer demands fall to their minimum level during the night, rise steadily during the 'morning peak' and reach a daily peak later in the day. That peak might be in the early afternoon – in hot climates, with a high air-conditioning demand – or in the early evening, when the industrial load is still high, and is augmented by additional transport, heating and lighting demands. The demand fluctuations over five winter days in western Denmark are shown in Figure 2.1. The maximum demand – at nearly 3,500 MW – is nearly twice the minimum demand, and the cycle is repeated every day, although demand levels are lower at weekends. The figure also shows the 'net' demand, taking into account the contribution from

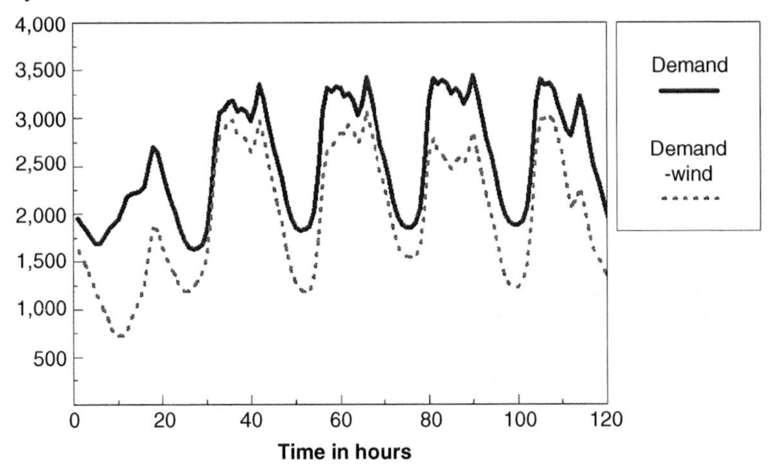


*Figure 2.1* System demand in west Denmark and the net demand, allowing for the contribution from wind

the wind plant. Although the capacity of the wind plant was around 1800 MW the consumer demands still predominate.

The electricity system in western Denmark is fairly small by international standards but it still benefits from the aggregation of consumer demands. A simple example shows how important this is. The minimum demand from a single house in western Europe is a few watts, the average is about 0.5 kW and the maximum is 5–10 kW – 10 to 20 times the average. If each UK household met its own maximum demand – 5 kW, say, 100 GW of plant would be needed for this sector alone. Aggregation smoothes variations in demand from all sectors so that, nationally, the maximum demand is around 60 GW, about 1.5 times the average demand. The same principle applies to generating plant. Although the probability of any particular generating unit being out of action at a particular time may be, for the sake of argument, around 1%, if system demand is being met by 100 units then the loss of a single one of those units will not cause undue problems. The probability of two or more units failing at the same time is very low.

The management of electricity systems is all about management of risks. One hundred percent reliability is not guaranteed, but a very high level of reliability can be attained by suitable operating strategies. The other important point is that risk management requires knowledge of the uncertainties in both generation output and consumer demand. The uncertainties in generation output can be established once the failure probabilities of the various types of plant are known. The treatment of consumer demands is somewhat similar – most of the fluctuations in the Danish demand, shown in the figure, can be predicted with reasonable accuracy. However, industrial, consumer and domestic demands do not always follow expectations. Television programmes, for example, may turn out to be more (or less) popular than expected, so that the surges in domestic demand when they end are more (or less) than predicted.

## Coping with the unexpected

To deal with unexpected mismatches between supply and demand, most System Operators place contracts with suppliers of various 'ancillary services' that ensure system frequency and voltage are maintained within statutory limits. These are called on as required on timescales from seconds (frequency response) to hours (standing reserve). Table 2.1 summarises the average requirements for the British network (Ofgem, 2004). The figures are also normalised with respect to the peak demand,

**Table 2.1** **Summary of ancillary service requirements in Great Britain**

| Type | Description | Timescale | GB holding (MW) | Holding/ peak demand |
|---|---|---|---|---|
| Frequency response | Automatically increases output if frequency falls, and vice versa | Seconds to minutes | 1,000 | 1.6% |
| Regulating reserve | Usually 'spinning'; adjusts output at request of System Operator | ~ 2 min to 1 hour | 1,900 | 3.1% |
| Standing reserve | Not necessarily synchronised | 20 mins upwards | 900 | 1.5% |

*Note that nomenclature for these services is not uniform across all electricity jurisdictions.*

and similar requirements are found in many other electricity networks. However, the advantages of aggregation mean that the requirements in smaller systems may be proportionately larger. The levels of reserve required at any given time depend partly on uncertainties in the predictions of demand, but also on the need to deal with the sudden loss of substantial amounts of generation, either due to power station faults or the loss of transmission circuits. In Britain, for example, key criteria are possible loss of one circuit of the cross-Channel link (1,000 MW), or of Sizewell B nuclear power station (1,200 MW).

It may be noted that the table makes no reference to demand-side management or to pumped storage. These are both specific types of regulating reserve and will be contracted by System Operators provided they can meet the technical specification for the particular type of reserve – and are economic. Demand-side management in Great Britain in 2003/04 accounted for around 20% of reserve costs. The total cost of these services in Great Britain for the financial year 2005/06 was estimated to be £257 million, or around £0.8/MWh of all electricity generated (€1.2/MWh, approximately).

The key issue that needs to be established is the extent to which introduction of wind energy in an electricity network increases reserve holdings, and the consequential cost increases. In doing this it is vital to note:

- Uncertainty margins do not add arithmetically, and
- All uncertainty margins come with a spectrum of probabilities

Although uncertainty increases with the time horizon, broadly speaking, the costs of the appropriate reserve decrease. 'Standing reserve', for example, may cost around £1/MW-h, but frequency response plant may cost up to £5/MW-h, or more.

## Wind characteristics

Just as combining the demands from more and more consumers smoothes the electricity demand on the network, so bringing increasing numbers of wind farms together smoothes the overall output. As the amount of wind energy on a network increases, the wind farms are likely to be spread more widely over the country. The greater the distances involved, the greater the smoothing, as the correlation between wind speeds from different sites decreases with distance.

The way that increased geographical spread reduces wind fluctuations is illustrated in Figure 2.2, which compares the output of a 5 MW wind farm over a 48-hour period with the output from all the 2,400 MW of wind in western Denmark. The output from the single wind farm is clearly more 'jagged'.

A more useful way of presenting the data on wind fluctuations is shown in Figure 2.3. (Milborrow, 2001). This compares the percentage of time that total wind power changes within one hour that have been recorded in western Denmark, over a typical year, with the corresponding power changes recorded from a 5 MW wind farm.

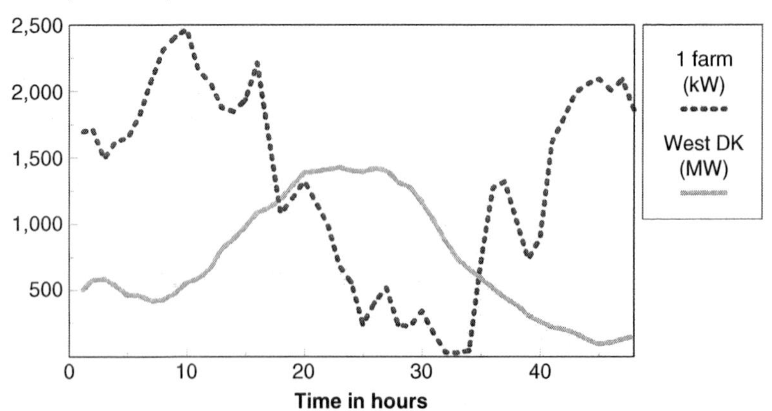


*Figure 2.2* Typical variations in output from a single wind farm and for all the wind in western Denmark

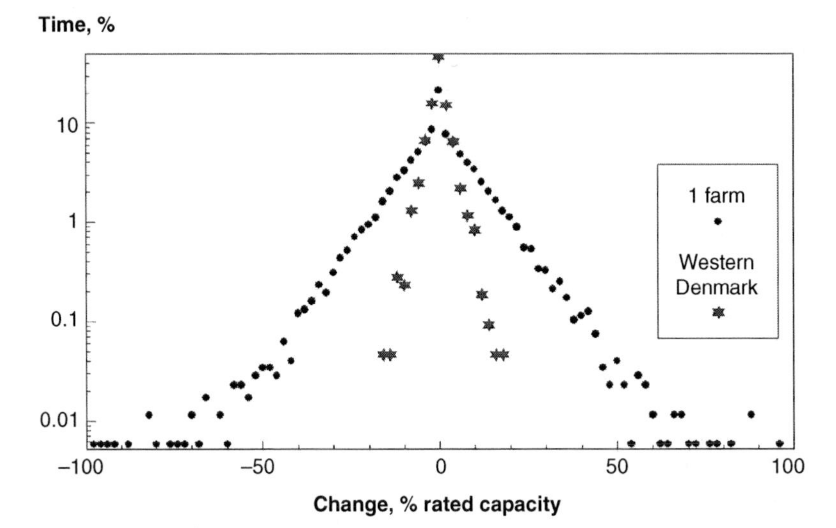


*Figure 2.3* Changes in the output of a single wind farm, between two successive hours, compared with the corresponding changes in the output of all the wind plant in western Denmark

The data from western Denmark in Figure 2.3 show that, for 78% of the time, the power changes within one hour by less than plus or –3% of its initial value. The corresponding figure for a single wind farm is 38% of the time. At the other end of the scale, the output from a single wind farm may, very occasionally, change by 100% within an hour. In western Denmark, on the other hand, there were never any changes greater than 18%. A very similar pattern of fluctuations has been observed in Germany (ISET, 2005).

The fact that the output from distributed wind farm stays within plus or –3% of its 'hour zero' value for 78% of time is tremendously important in the context of managing its impacts on an electricity network. Even looking four hours ahead, the power output will be within plus or –3% of its initial value for 42% of the time. Data of this kind provide System Operators with the information they need to manage wind energy and this is discussed in the next section.

## Operating electricity networks with wind

System operators simply cannot detect the variations in output from small amounts of wind plant since they are swamped by numerous other fluctuations of similar magnitude. When gale force winds cause a

5 MW wind farm to shut down, it is unlikely that all the turbines will cut out at the same time; output will be curtailed in over several minutes, but a German 'ICE' train, or an Anglo-French Eurostar train which shuts off power and coasts may 'switch off' a 5 MW load – or more – in a matter of seconds.

The wind power fluctuations, in themselves, are not the crucial factor in determining how easy (or difficult) it is for System Operators to cope with wind energy. What also matters is

- Whether the addition of wind results means that System Operators need to cope with more rapid changes in supply/demand balance, and
- The additional uncertainty in the supply/demand balance.

## Changes in demand patterns

Figure 2.4, using actual data from western Denmark, addresses the first point. The figure shows the changes in demand on the power system between successive hours.

It shows, for example, that for 14% of the time there is very little change in demand between two successive hours. At the other end of

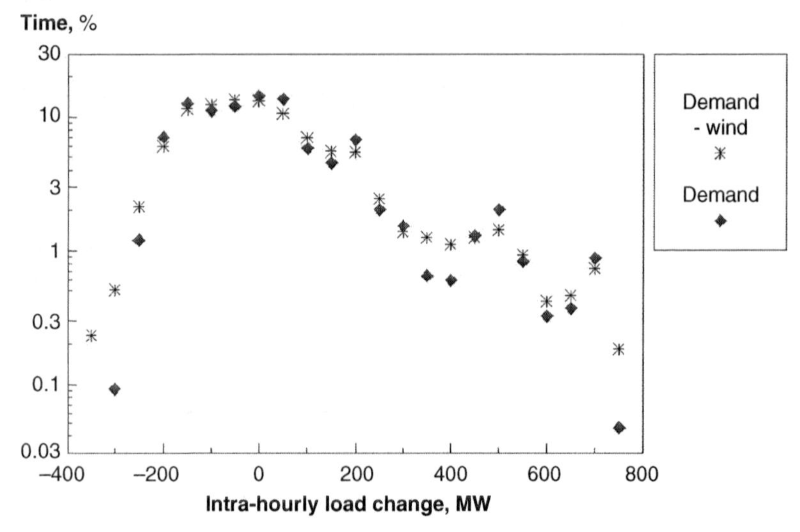


*Figure 2.4* Intra-hourly demand changes observed in western Denmark over the course of the typical year, compared with the corresponding changes with wind energy on the system

the spectrum, for 0.046% of the time (four hours) the power change between two successive hours is about 750 MW (roughly 20% of system peak demand). The maximum negative change is only about 300 MW. The curve labelled 'demand minus wind' is what the system operator saw in 2001. It is the demand, net of the wind energy generation. Although there are differences between the two curves, there are no ugly surprises. The number of 750 MW changes goes up from four to about 16 hours a year, and the maximum negative change increases slightly – to about 350 MW – again for about 16 hours a year.

## The extra uncertainty due to wind

The data shown in Figure 2.3 provide a basis for estimating the additional uncertainty when wind is added to a network, and hence the requirements for extra reserve. The standard deviation of the one-hour uncertainty (and any other time-ahead values needed by the System Operator) of the wind power can readily be calculated and typical results are shown in Table 2.2 (Milborrow, 2001). The table includes data for a single wind farm, providing further confirmation of the beneficial effects of countrywide smoothing.

Modest amounts of wind cause few problems (or costs) for System Operators since the extra uncertainty imposed on a System Operator by wind energy is not equal to the uncertainty of the wind generation, but to the combined uncertainty of wind, demand and thermal generation. This combined uncertainty is determined from a 'sum of squares' calculation:

$$\sigma^2 \text{ (total)} = \sigma^2 \text{ (demand/generation)} + \sigma^2 \text{ (wind)}$$

The uncertainty in the supply/demand balance depends, in turn, on the plant reliability statistics and on demand uncertainty. In a typical

***Table 2.2*** **Standard deviation of wind power fluctuations (per cent of wind capacity)**

| **Lead time, hr** | **1** | **4** |
|---|---|---|
| **Nation-wide** | | |
| NGC (2001) (UK system operator) | 3.1 | 6.0 (at 3.5 hr) |
| West Denmark | 3 | 10 |
| **Single wind farm** | 11.8 | 20.8 |

power system, on average, four hours ahead, both are roughly equal to about 1.3% of system demand, so the combined uncertainty is about 1.8% of system demand. The corresponding standard deviation in the uncertainty of wind generation, four hours ahead, is around 6% (Table 2) and so the extra reserve needs can be derived. These are often based on the assumption that the system operator schedules three times the reserve corresponding to the standard deviation (3 × 1 standard deviation) of the uncertainty, on the basis that three standard deviations capture 99% of all likely scenarios. A simple analysis suggests that the extra reserve needs will be 2.4% of the wind capacity when the latter is 5% of the peak demand, rising to 4.6% of wind capacity at 10% of peak demand, and just under 8% of wind capacity at 20% of peak demand. These results compare well with analyses by utilities.

Calculations of this type are made on various timescales to determine the total needs for extra reserve. One hour ahead the uncertainties are less – but the shorter-term reserve costs more.

## Costs of extra reserve

Estimates of extra reserve capacity can be misleading since it is not always clear whether it is short-term reserve, medium-term, frequency response plant, or the sum of all three. A better guide to the impacts of variable renewable sources is the total costs of all the extra reserve. Broadly speaking, the characteristics of many electricity systems are similar, together with the costs of reserve, so estimates of the extra reserve needed to cope with wind energy are also similar. For the UK, the System Operator (Dale, 2002) has suggested that the extra balancing costs for wind at the 10% level would be £40 million per annum (£2.4/MWh of wind); at the 20% level wind would increase balancing costs by around £200 million p.a. (£2.8/MWh of wind). Data from several other studies have yielded similar results, shown in Figure 2.5. (Most of the studies are American, so values have been left in dollars). With 5% wind, the extra costs are within the range $1.7–3/MWh, and with 10% wind the range is $3–5/MWh.

The prices that generators charge system operators for reserve take into account the lower thermal efficiency of generating sets when running at part load. (The sets that provide fast reserve usually operate at part load so that they can supply additional power when requested). It is sometimes argued that the additional $CO_2$ emissions associated with part load operation of coal or gas-fired plant offset the $CO_2$ savings associated with wind energy by a significant amount. This is

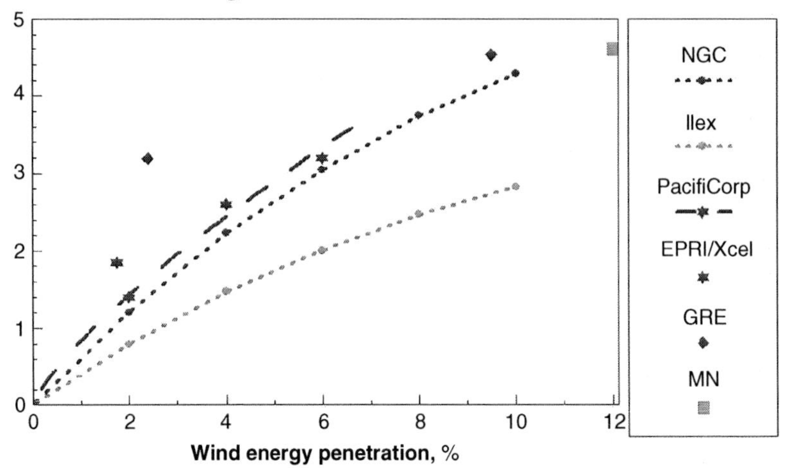


*Figure 2.5* Estimates of the cost of extra balancing needed for wind in US $/MWh

The data shown in Figure 2.5 come from the following sources:
** NGC: (For England and Wales), Dale (2002)
** Ilex: A UK study by Ilex Consulting (2002)
** PacifiCorp: An analysis for the Power Utility PacifiCorp (Coatney, 2003)
** EPRI/Xcel: A 'Case Study' for Xcel Energy (Electrotek Concepts, 2003)
** GRE: An analysis for Great River Energy (Seck, 2003)
** MN: An analysis for the State of Minnesota (EnerNex Corporation, 2004).

*Note*: as the majority of these studies were American, prices have been left in US dollars, and the UK data converted on the basis that £1 = $1.8 at the time of writing. In the next section, European data are quoted in Euro, where 1 Euro = £0.7. For completeness, it may be noted that 1 Euro = $1.26 (2006).

incorrect. In the first place, the amount of additional part-loaded plant required for wind energy is small and, secondly, the changes in thermal efficiency are modest (around 10%, at most). Dale *et al.* (2004) estimated that a 20% contribution to electricity supplies would reduce emissions savings due to wind by about 1%, compared to the theoretical value. In other words, if wind displaces coal, emitting around 900 g/kWh of carbon dioxide, 20% wind would save around 890 g of carbon dioxide per unit of wind energy generated.

## The case of Germany

The extra costs of reserve in Britain and a number of American electricity jurisdictions are similar (Figure 2.5), but significantly higher figures are quoted for Germany. Although the characteristics of wind tend to be similar the world over and the same applies to electricity networks, there are some differences. In Germany, as well as in some other electricity jurisdictions, wind tends to be treated in the same way as gas or coal-fired plant and is required to forecast its output several hours ahead. If the plant schedules lack flexibility, it is quite likely that the output from the wind plant will change after the commitment is made. That may mean that balancing power must be purchased to make up any power deficits or, alternatively, surplus wind may need to be sold for a low price. The more flexibility that is built into plant scheduling, the more efficiently can the system the operated. This point is discussed later.

The other reason why balancing costs for wind in Germany tend to be higher than elsewhere is that wind speeds are lower. The average capacity factor of German wind is about 15%, which compares with 25% in Denmark and 30% in Britain. This means that the wind plant capacity needed to generate a given amount of energy in Germany is roughly twice the capacity needed in Britain. The magnitude of the power swings from the plant in Germany are therefore higher than those in Britain or Denmark. The additional uncertainty means that the system operator needs to schedule more reserve. The impact of capacity factor on extra reserve costs is illustrated in Figure 2.6. This shows that the additional costs associated with 5% wind energy penetration are around Euro 2/MWh in a region where the capacity factor is

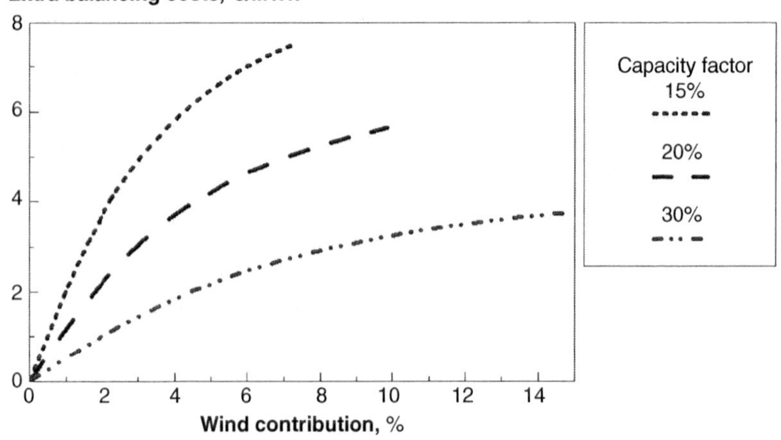


*Figure 2.6* The influence of capacity factor on extra balancing costs in Euro/MWh

30%, but nearly Euro 6/MWh in a region where the capacity factor is 15%. A very thorough analysis of the impacts of wind on the German network was completed in 2005 by a consortium of utilities and expert consultancies (DEWI *et al.*, 2005).

## The importance of wind prediction

Although there are differences in the way that electricity jurisdictions operate, there is a reasonable consensus on the savings that can be realised through good forecasting. These savings accrue since the uncertainties that System Operators face when handling wind energy are significantly reduced and this enables them to reduce the amount of extra reserve plant that is scheduled. Although the monetary savings depend on the costs of reserve, they are of the order £0.5–1/MWh at low wind energy penetrations (2–4%), rising to around £1.5–2/MWh with 10% wind energy. Considerable research is in progress, aimed at delivering better forecasts and a better understanding of the issues. (Kariniotakis *et al.*, 2003)

## Costs of reserve for other variable renewables

There are few estimates for the costs of reserve for solar, tidal and wave energy. The costs associated with assimilation of solar energy are likely to be similar to those of wind; it varies on relatively short timescales, but geographical diversity can smooth the fluctuations. The extra costs associated with tidal sources are likely to be quite small as they are predictable, although some of the large schemes could necessitate some extra reserve holdings.

Wave energy appears to be slightly less variable than wind energy – at least when its impacts can be diversified across a large region such as the British Isles. Sinden (2005) has suggested that the standard deviation of the hour-ahead variability is 2.6%, which compares with his figure of 3.2% for wind. This would reduce additional balancing costs by about 15%. Sinden also looked at the implications of operating with a combined wind and wave power portfolio and suggested that this would reduce extra balancing costs by 37%, compared with a wind-only portfolio.

## Storage

'Renewables need storage', claim their detractors – and sometimes their supporters as well. 'Storage can transform the economics of the intermittent renewables', proclaimed an august body recently.

Both quotes are very misleading; in the first place, it is only the variable sources of renewable energy that might benefit from storage and, even then, the use of storage only makes sense if its value is greater than its cost. On a small island, with wind or solar power as the only energy source, then the economics might stack up; elsewhere it isn't easy. Storage has no intrinsic merits for coupling with wind energy, as an early analysis by Farmer *et al.* (1980) made clear:

> ...there is no operational necessity in associating storage plant with wind-power generation, up to a wind output capacity of at least 20% of system peak demand.

A more recent study made the same point (Hirst, 2002):

> Storage may increase the value of intermittent generation. However, studies generally show that dedicated storage systems for renewables are not viable options for utilities because of added capital costs of current storage technologies. Storage can add flexibility and value to utility operations, but it should generally be a system-wide consideration, based on the merit of the storage system.

The idea, some argue, is that storage can be used to 'level the output' of the variable sources of renewable energy. Whether or not that output really needs to be levelled is a key issue. Storage may well be beneficial – perhaps even cost-effective – as part of the strategy for operating the complete system, but isolating particular generating plant simply rarely makes economic sense.

Another way of looking at the issue is to examine the way in which consumer demand and wind generation behave. Analysis of data from western Denmark suggests that, for 52% of the time, wind and demand are either rising together, or falling together. There is little point in sending wind energy to a store if it is helping the System Operator to cope with rising demand. Conversely, there is little point in drawing down from a store when wind output is falling, if the overall demand is falling.

## Storage options

Of the various storage technologies, pumped storage is the most firmly established, with around 22,000 MW installed in the US and over 2,000 MW in the UK. It is proven technology – the turbines used are similar to those found in Hydro Electric plant, but they are reversible and pump the water uphill when necessary and are fast. The other advantage is that they can deliver power for minutes or even hours at a time; most other technologies struggle to deliver (or absorb) power for long periods.

In terms of installed capacity, compressed air storage follows next – but some way behind – with US capacity around 100 MW. As the title implies, compressed air is pumped into caverns and then released back – usually into gas turbines. Research is in progress in Germany and elsewhere, apart from the US. 'Flow batteries' – where chemical reactions in electrolytes take the place of the water in a pumped storage system – show some promise. The 'Regenesys' technology, developed by National Power, falls into this category. Although its new owner has discontinued the work – perhaps confirming that it is difficult to make ends meet – other, similar, systems are being developed in Japan and elsewhere.

## It's all down to economics

Several studies have suggested that the target cost for storage systems to be economic is around £500–800/kW, e.g. Strbac and Black (2004). It is difficult to be precise, as it depends what revenues can be realised from various 'reserve' services. That assumes a pretty intensive use for the system, remembering that the maximum load factor of a storage system cannot exceed about 35–40% – it spends half its time charging, and the overall efficiency will be around 80%, at the very most. Conventional batteries can meet this cost target – but their storage potential is limited. Pumped storage systems in very favourable locations can meet the target. Compressed air storage might be able to meet the target – but exploitation is so far rather limited. A storage system used solely to 'level the load' of wind or solar power plant would almost certainly have an even lower load factor than 35% – so its energy delivery cost would be high. Given that the difference in value between 'firm' and 'non-firm' power rarely exceeds about £5/MWh – about one-tenth of the delivery cost, this shows why storage for variable renewables is unlikely to make economic sense. It may make sense for island situations, however, where the economics are completely different.

## Demand-side management

Demand-side management has a similar role to storage – it may be the most economical way for System Operators to provide reserve. It is an area of increasing interest and ideas for remote control of non-essential consumer loads are being investigated. As with storage, there may be opportunities for links with wind energy developments – depending on the economics. It may be a viable way of increasing the amount of wind generation, which can be accepted onto a weak network.

## Total costs and benefits of variable renewables

In a previous section it was observed that the value of 'firm power' was only slightly higher than the value of 'non-firm power'. That is because capital repayments for gas-fired generating plant (the usual source of 'firm power') make only a small contribution to their generating costs, compared with the costs of fuel.

Nevertheless, when assessing the overall financial viability of renewable energy sources, their ability to substitute for other plant needs to be determined, as the total value of variable renewables is:

(fuel saving value) + (capacity saving value) – (costs associated with intermittency)

If this total value exceeds the generating cost of variable renewable, electricity consumers in the network concerned will pay less for their electricity, but if the reverse is true, they will pay more for their electricity.

The capacity saving value of variable renewables depends on their capacity credit. This is a measure of the ability of the plant to contribute to the peak demands of a power system. Capacity credit here is defined as the ratio (capacity of thermal plant displaced)/(rated output of wind plant).

## Capacity credit

Numerous utility studies have concluded that wind can displace thermal plant. The capacity credit of wind in northern Europe is roughly equal to the capacity factor in the winter quarter. Results from ten European studies are compared in Figure 2.7, showing credits declining from 20–40% at low wind penetrations to 10–20% with 15% wind. It should be noted that the values of capacity credit depend on the capacity factor of the wind plant.

To facilitate comparisons between the studies, Figure 2.8 compares normalised values (with respect to the capacity factor) of capacity

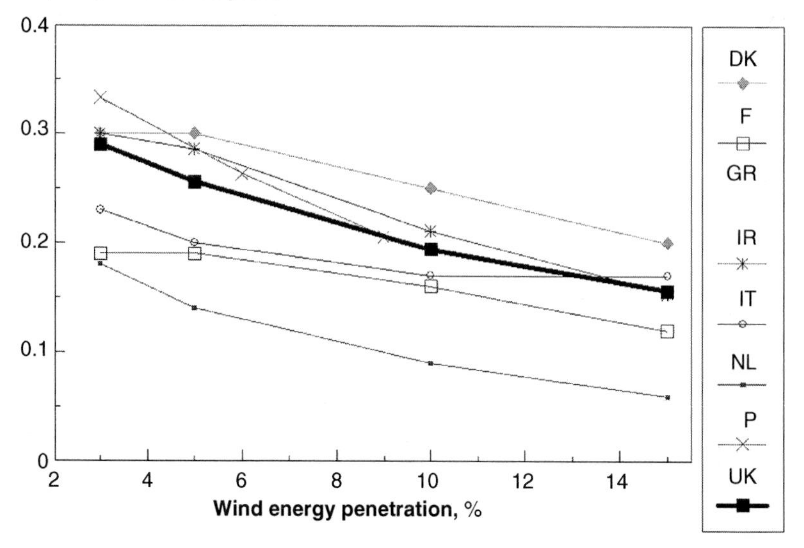


*Figure 2.7* Comparison of results from ten utility studies of capacity credit
*Source*: see Holt *et al*. (1990).

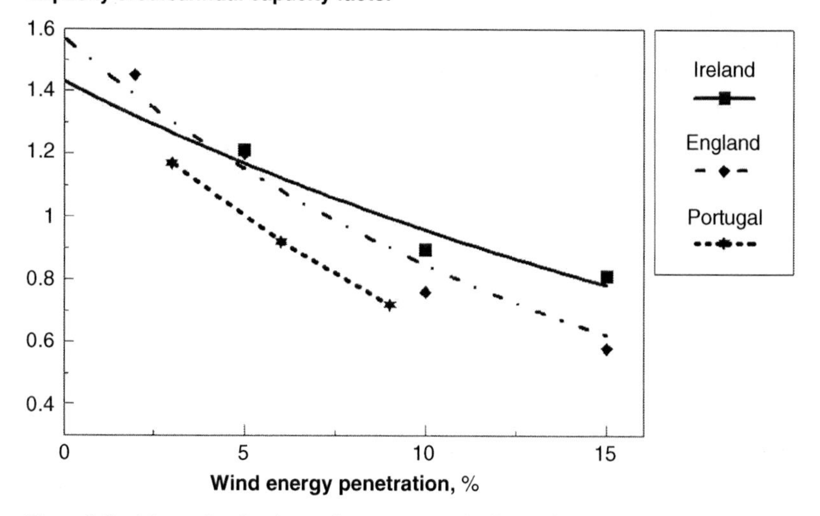


*Figure 2.8* Normalised values of capacity credit from three European studies

credit – drawn from the data of Figure 2.7 – and shows a good measure of agreement between three utilities. With modest contributions of wind energy the capacity credits are about 40% greater than the annual capacity factor of the wind plant and so, if the average wind capacity factor across a country is 35%, then 1,000 MW of wind would displace 490 MW of thermal plant. At higher wind energy penetrations the capacity credit declines, reaching 60–80% of the capacity factor at 15% penetration.

Provided that wind power and system demand are statistically independent, it can be shown that the capacity credit – at low penetrations – is roughly equal to the capacity factor of the wind plant. On the other hand, if wind and demand are well correlated, that increases the capacity credit of wind plant, and one Californian study suggested that the capacity credit of one of the early machines was about 80% of its rated power (Pacific Gas and Electric Company, 1989). At the other end of the spectrum, when wind and demand are negatively correlated, capacity credit is much lower. A study for the North West Power Pool suggested a value for that region of only 5% (Flaim and Hock, 1984).

## Capacity credit of technologies other than wind

### Tidal barrage

Although tidal energy is predictable, which means that management of the variability is subject to less uncertainty than for wind, it is actually less likely to contribute to peak power demands, which strongly influence the value of capacity credit and so the capacity credit of the Severn barrage has been assessed as 15.3% (Goddard, 1988). This, it may be noted, is lower than the capacity factor, which at the time was estimated to be 20%. The value may, however, be a function of the large size of the installation, as the capacity credit of the Mersey scheme (500 to 600 MW) was quoted as being around 28%. Similar considerations may be expected to apply to tidal stream technology.

### Photovoltaics

The capacity credit of solar PV installations in northern Europe is low. The contribution of the power source at times of system peak demand plays a dominant role in the calculations. System peak demands tend to occur around 17:30 on January weekdays, at which time the contribution from PV may be confidently taken as zero. However, studies in California have concluded that PV can realise significant capacity credits, as the system peak demands occur during the summer and there is a good correlation between solar power and air conditioning demand. Similar remarks apply to Greece. It should also be noted that the capacity factor of PV installations is significantly lower than those of wind plant.

## The value of capacity

Once the capacity credit has been determined, the 'value of capacity' can then be estimated. The reference new generation technology is usually combined cycle gas turbines, costing around £450/kW, and the annual capital replacement cost of such plant is about £45/kW, although this depends on the annual cost of capital. At low wind energy penetrations the value associated with the capacity credit is around 0.5 p/kWh. This is quite modest and explains why the concept of 'levelling the output' of wind farms is not easily achievable, as it is difficult, if not impossible, to purchase electricity at this price.

## Results of analyses of total costs and benefits

Recent examples of studies of the costs (or savings) accruing from the introduction of wind energy that take all the relevant factors into account have suggested the extra costs (if any) are modest. Dale *et al.* (2004) suggested that the extra cost to the UK electricity consumer of providing 20% of supplies from wind energy would be about £3/MWh. For Pennsylvania Black and Veatch (2004) suggested that a 10% renewables portfolio by 2015 would increase costs by $0.4/MWh. Wind contributed about 65% of the renewables mix. In the light of recent gas price rises the UK estimate is now very pessimistic, since, as the price of gas goes up, so the 'fuel saving value' of wind energy also goes up. The paper by Dale *et al.* (2004) used a UK delivered gas price of about 19 p/therm, but the average price in 2005 was nearly twice that figure. It is now difficult to quote future gas prices with any certainty, and a question mark also hangs over estimates of future wind plant costs, due to increases in wind turbine prices in 2005/6.

To deal with this uncertainty Milborrow (2006) has calculated 'break even' costs; for 10% wind these range from about Euro 800/kW for a gas price of €0.5/therm, to just under Euro 1,400/kW for a gas price of Euro 1/therm. The analysis assumed that equal quantities of onshore and offshore wind are installed, and that installed costs offshore are 50% higher than the onshore costs quoted. It was also assumed that gas-fired plant attracts a $CO_2$ cost of Euro 24/tonne. Although the analysis was carried out using parameters for the British electricity network, very similar results are likely for other systems.

Another way of assessing the likely costs of installing significant quantities of wind energy is to calculate the 'variability premium'. This takes into account both the extra balancing costs and the increased generating costs of the thermal plant. The latter costs rise as the thermal plant operates at progressively lower load factors as the amount of wind energy on the system increases. The Danish system operator, Energinet, (Pedersen *et al.*, 2006) suggested that the variability premium was Euro 10/MWh with 10% wind, rising to just over Euro 12/MWh with 20% wind, and €15/MWh with 60% wind, as shown in Figure 2.9. A similar analysis for Great Britain, using the methodology of Dale *et al.* (2004), derived slightly lower figures – Euro 8/MWh with 20% wind, rising to just under Euro 11/MWh with 60% wind (Milborrow, 2006).

The differences between the Danish and British results mostly stem from the lower capacity credit assigned to wind in Denmark.

## Trading arrangements

Electricity trading in many places, including Great Britain, California and Denmark, require each electricity supplier to achieve a short-term balance, typically 1–4 hours. This tends to ignore the benefits of aggregation. Shortfalls in power, after 'gate closure', (when the likely power output

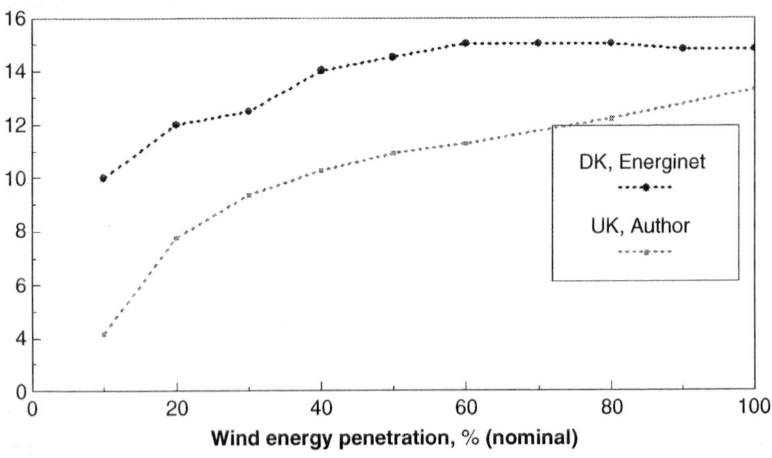


*Figure 2.9* 'Costs of variability' for wind in Britain and Denmark in Euro/MWh

must be quantified) are made good at a 'top up price' and surpluses are sold at 'spill price'. This may be same as the 'top-up' price, or lower.

In California, exemption from these 'imbalance' penalties has been agreed, along with other measures to ease the assimilation of wind energy. There are moves to extend this provision across the whole of the US.

## Transmission constraints

Transmission limitations may sometimes impose constraints on the assimilation of wind. In the UK and Germany, for example, concentrations of wind in the north – where wind speeds are higher than those further south – have increased the north-south power flows, which are already substantial. Substantial reinforcement costs for the transmission networks may be needed, depending on the timing and location of new renewable generation. In Scotland, studies carried out by the network operators suggest that 6,000 MW of new and renewable capacity in Scotland might trigger reinforcement costs up to £1,500 million.

## Local issues

Local issues are complex and vary both regionally and locally. Concentrations of embedded generation can increase distribution losses in rural areas where demand is low, although modest amounts may reduce losses. A study of a ten-machine, 4 MW wind farm connected into an 11 kV system in Cornwall, provided valuable information on local issues (South Western Electricity, 1994). The study concluded 'the wind farm caused surprisingly little disturbance to the network or its consumers'. In particular:

- Voltage dips on start-up were within prescribed limits.
- There were no problems with flicker during any operating conditions.
- During periods of low local load, the output from the farm was fed 'backwards' through the distribution network, but no problems were reported.
- Reduced activity of the automatic tap-changers at the adjacent 33/11 kV transformers was significant and would lead to lower maintenance costs.

Large current fluctuations were sometimes observed although there was a possibility that these originated elsewhere within the distribution system.

## Concluding discussion

Much of the early work on wind integration involved simulation studies and the results are now being tested against actual operational data. These comparisons strongly indicate that some of the early estimates of wind fluctuations were on the high side, and so some early estimates of the 'costs of intermittency' are likely to be pessimistic.

The overwhelming consensus, from the studies cited in this paper and from a wide-ranging review of the relevant literature, worldwide (Gross *et al.*, 2006), is that there are no major technical barriers to the implementation of dispersed intermittent generating systems connected to the network. The costs of managing the additional uncertainty associated with the variability are in the range £1.6–2.4/MWh with 10% wind energy, rising to £2–3/MWh with 20% wind energy.

A number of factors are likely to reduce the impacts of intermittency. Better wind prediction is a key issue which will reduce the uncertainty – and hence the cost – of absorbing wind energy and research is under way in both America and Europe to develop better techniques. The importance of this work will increase in the future, as the proportion of wind energy in electricity networks increases. Although the monetary savings depend on the costs of reserve, they are of the order £0.5–1/MWh of wind at low wind energy penetrations (2–4%), rising to around £1.2–1.7/MWh with 10% wind energy.

Market mechanisms that require individual suppliers to 'balance their own positions' can introduce cost penalties for wind energy that are not cost reflective. It is difficult to quantify typical monetary levels, as the relevant prices in the market are rarely cost-reflective and vary with time, and across electricity jurisdictions. The penalties incurred by wind are not due to any real technical problem, but to the vagaries of a particular set of market rules. However, there is a growing trend, particularly in US, towards exempting wind from balancing market penalties, often by averaging imbalances over a month, and concentrating on providing System Operators with good forecasting tools – for which the wind plant operators pay a levy. The magnitude of the levy roughly corresponds to the costs of the extra balancing that is required – from the standpoint of the system as a whole.

Considerable worldwide interest in the potential of demand-side management techniques has the potential of reducing balancing costs for System Operators and so, as a side effect, reducing the additional costs of intermittent renewables. Looking to the future, there is now considerable interest in exploring the possibilities of high penetrations of wind energy into electricity networks. The cost implications at the higher penetration levels are inevitably somewhat less certain, but do not appear excessive.

Given the recent increases in the price of gas and the strong downward trend in wind energy prices that has been evident for some time, the overall conclusion that can be drawn is that the future prospects for wind energy are bright.

## References

Black and Veatch Corporation (2004) Economic impact of renewable energy in Pennsylvania.

Coatney, T. (2003) Modelling wind energy integration costs. Utility Wind Interest Group, Technical Wind Workshop, Seattle, 23 October.

Dale, L. (2002) Indicative Costs of Wind. University of Oxford, Environmental Change Institute. 'Renewable Energy and Intermittency' Workshop, 29th November. http://www.eci.ox.ac.uk/lowercf/renewables/intro.html

Dale, L., Milborrow, D., Slark, R. and Strbac, G. (2004) Total cost estimates for large-scale wind scenarios in UK. Energy Policy, 32, 1949–56. First published in Power UK, issue 109, under the title: 'A shift to wind is not unfeasible'.

DEWI, E.ON Netz, EWI, RWE Transport Grid and VE Transmission (2005) Planning of the Grid Integration of Wind Energy in Germany Onshore and Offshore up to the Year 2020. http://www.eon-netz.com/Ressources/downloads/dena-Summary-Consortium-English.pdf

Electrotek Concepts (2003) Characterising the impacts of significant wind generation facilities on bulk power system operations planning.

EnerNex Corporation (2004) Wind Integration Study for Xcel Energy and the Minnesota Department of Commerce.

Farmer, E.D., Newman, V.G. and Ashmole, P.H. 1980. Economic and operational implications of a complex of wind-driven power generators on a power system. IEE Proc A, Vol. 127, No. 5.

Flaim, T. and Hock, S. (1984) Wind energy systems for electric utilities: a synthesis of value studies. Solar Energy Research Institute, SERI/TR-211-231.

Goddard, S.C. (1988). Comparison of non-fossil options to Hinkley Point 'C'. Proof of evidence for Public Inquiry, CEGB.

Gross, R., Hept'onstall, P., Anderson, D., Green, T., Leach, M. and Skea, J. (2006) The costs and impacts of intermittency. Imperial College, London.

Hirst, E. (2002) Integrating wind energy with the BPA power system: preliminary study. Consulting in Electric – Industry Restructuring, Oak Ridge, Tennessee.

Holt, J.S., Milborrow, D.J. and Thorpe, A. (1990) Assessment of the impact of wind energy on the CEGB system. Commission of the European Communities, Brussels. The Commission supported similar studies of other European utilities.

Ilex Energy Consulting Ltd and UMIST (2002) Quantifying the system costs of additional renewables in 2020. Report commissioned by DTI, [The 'SCAR' report].

ISET (Institut fur Solare Energieversorgungstechnik) (2005) Wind Energy Report Germany 2004.

Kariniotakis, G. *et al.* (29 authors) (2003) ANEMOS: development of a next generation wind power forecasting system for the large-scale integration of onshore and offshore wind farms. European Wind Energy Conference, Madrid.
Milborrow, D.J. (2001) Penalties for intermittent sources of energy. Working paper for UK Energy Review, www.pm.gov.uk/files/pdf/Milborrow.pdf
Milborrow, D.J. (2006) Nuclear suddenly the competitor to beat. Windpower Monthly, January.
Milborrow, D.J. (2006) The economic myths of high wind penetration. Windpower Monthly, September.
Ofgem (2004) NGC System Operator incentive scheme from April 2005. Initial proposals.
Pacific Gas and Electric Company (1989). Solano MOD-2 wind turbine operating experience through 1988. Electric Power Research Institute, GS-6567.
Pedersen, J., Eriksen, P.B. and Orths (2006) Market impacts of large-scale system integration of wind power. European Wind Energy Association Conference, Athens.
Seck, T. (2003) GRE wind integration study. Utility Wind Interest Group, Technical Wind Workshop, Seattle, 23 October.
Sinden, G. (2005) Diversified renewable energy portfolios for the UK. British Wind Energy Association, 27th annual conference, Cardiff.
Strbac, G. and Black, M. (2004) Future value of storage in the UK. Manchester Centre For Electrical Energy.
Sustainable Development Commission (2005) Wind power in the UK.
South Western Electricity plc, 1994. Interaction of Delabole wind farm and South Western Electricity's Distribution system. ETSU report W/33/00266/REP.

# 3

# Biomass – Greening the Transport Sector

*Jonathan Scurlock*

## Introduction

Transport fuels supply around one third of total energy use in the UK, and nearly a quarter of national carbon dioxide emissions. Until recently, there seemed little that could be done to curb the growing demand for transport, and there were few practicable options for cleaner delivery of transport services. However, the British government's Renewable Transport Fuels Obligation (RTFO) is now promising verifiable cuts in carbon emissions together with improved security of supply, based upon substituting a rising proportion of conventional fuels with biofuels derived from agricultural crops such as oil seed rape, palm oil, wheat and sugar. Biomass fuels or biofuels are defined as those made from any kind of plant matter (agricultural or forestry products, by-products, residues and wastes), whereas most transport fuels are presently derived from oil (with a small contribution from other fossil fuel resources such as coal and natural gas).

Are biofuels a viable option? What about other novel technologies such as hydrogen-powered fuel cell vehicles? Can biofuels ever supply more than a few per cent of our transport energy needs? Is this a new opportunity for domestic farmers and biofuel growers in developing countries, or will it only put further pressure on land use?

Britain is only one of many industrial economies developing biofuels. The European Commission is pressing ahead with its EU Biofuels Strategy, which prepares the way for large-scale uptake of both domestic and imported biofuels, and the USA is in the throes of a large-scale expansion of ethanol production from maize, with many new refineries owned by mid-western farmer cooperatives, offering a lifeline to economic security. This may be all very well in the United States and those parts of

Europe where the population density is relatively low, but does it make sense in the UK, where land is more of a premium resource? If Britain expands the use of biofuels, how much can we grow ourselves, or will we have to rely upon substantial imports? Is it reasonable or indeed sustainable to rely upon developing countries to supply the necessary feedstocks?

This chapter will explore some of the background to Britain's biofuel revolution and the evidence underpinning the RTFO, including the feasibility of meeting a substantial fraction of UK transport needs from a combination of domestic and imported biomass feedstocks. Of all the renewable energy options, the UK's biomass resource is probably second only to its wind resource, where biomass is defined in its widest sense, as plant-derived matter used for heating, power generation or conversion to transport fuels. However, bringing most kinds of biomass to market has not been easy – trading in wood fuels for power generation and space heating is still in its infancy. It is hoped that integration of the initial transport biofuel options into existing agricultural supply chains will enable them to be implemented relatively quickly. Nevertheless, there is potential for conflict between the competing demands for different uses of biomass from an ultimately limited resource base.

The following discussion is concerned principally with the resource base and sustainability of biomass-derived transport fuels. Other applications of the wide range of biomass feedstocks, in the heating and electric power generation sectors, are considered only in the context of competition for limited resources. Apart from climate change, the transport sector needs to address other aspects of sustainability, such as the local human health impacts of vehicle emissions, but these too receive only brief mention here.

In the end, the contribution that biomass makes to future transport energy needs will depend also upon the level of improved energy efficiency that we adopt, whether in specific vehicle technologies or transport systems in general. That it turn links into the wider issue of transport policy and in particular attempts to limit emissions via a shift to public transport and the adoption of new guidelines on local/regional planning and economic development.

This is a very large agenda. The initial role of liquid biofuels is to provide a practical first step towards reducing emissions from transport, principally road transport. It is hoped that they will also act as a transitional strategy, laying the basis for a more substantial move into the sustainable provision of transport services. As this chapter illustrates, the next steps beyond these 'first generation' transport biofuels are not yet a foregone conclusion.

## The starting point for biofuels

In 2005, Britain consumed about 53 million tonnes (Mt) of transport fuels with an energy content of 2,200 petajoules, which accounted for about one-third of total energy use (DTI, 2006). Carbon dioxide emissions from road transport (33 Mt of carbon equivalent) comprised more than one-fifth of the UK total (DEFRA, 2006). Annual transport fuel consumption is dominated by petrol (18.7 Mt) and road diesel (19.5 Mt). Aviation fuel consumption (as supplied by UK airports) is estimated at 12.5 Mt, but this is harder to attribute to the UK in particular, since airlines may buy fuel in many different countries of operation. Recent estimates suggest that aircraft produce about 5.5% of UK carbon emissions, and that aviation emissions doubled during the 1990s (ECI, 2006). A much smaller part of transport fuel consumption is attributed to rail (0.8 Mt) and water transport (0.9 Mt). Again, the latter has an international dimension, and much consumption of marine fuel oil is not included in UK statistics (world consumption is around 150 Mt, of which Europe comprises about 40 Mt). For the year 2005, Britain reported under the European Union Biofuels Directive that just 0.3% of road transport fuels were of biomass origin, i.e. produced from plant (vegetable) matter that recycles carbon dioxide from and to the atmosphere. The rest was produced mostly from petroleum, and so the vast majority of UK transport is powered by fossil fuels that contribute carbon dioxide permanently to the atmosphere.

Despite growing concerns about climate change and 'peak oil', total oil consumption in the UK is still increasing – it rose by 2% in 2005 to 82 million tonnes of oil equivalent. The majority of this, about 80%, was consumed in the transport sector. Conversely, about 97% of UK transport energy is derived from petroleum. Sustainable energy options for transport in Britain appear to be severely constrained by the degree of lock-in to technology and energy distribution infrastructure. The supply chain for both fuels and vehicles is managed mostly by large transnational companies whose turnover dwarfs many individual nation states. Thus there is a range of both strategic and environmental reasons for reducing the carbon intensity of Britain's road transport sector.

A similar degree of dependence and lock-in is found in many other OECD countries and some parts of the developing world (UNDP, 2004). In the USA, petroleum products (gasoline, diesel and jet fuel) provide virtually all of the energy consumed in the transport sector, and transport is the greatest single use of petroleum, at about 67% of total oil consumption in 2004 (EIA, 2006). Japan's transport sector comprises about 25% of total energy use, although its overall petroleum dependence has decreased from around 77% of primary energy in 1973 to 49% in 2001 (ANRE, 2006). Oil provides about 33% of commercial primary energy in India, and the country imports over 70% of its oil (MOSPI, 2006). In both Japan and India, transport is about 98% dependent upon petroleum products (IEA, 2006a). Overall, in 2001, transport accounted for 57% of primary oil demand in OECD countries and nearly 40% in non-OECD countries (IEA, 2003a, 2003b).

Until recently, the UK lacked policy measures specifically designed to replace fossil fuels in transport with renewable fuels. Government policy on renewable energy was focused almost exclusively on the electricity generating industry, which accounts for just one-third of primary energy use and about one-third of carbon emissions. In response to the EU Biofuels Directive of 2003, which set indicative targets for 2005 and 2010 (see below), the British government announced a Renewable Transport Fuels Obligation (RTFO) in November 2005. Due to come into effect in April 2008, the RTFO will require at first that 2.5% (by energy content) of road vehicle fuel in the period 2008–09 is from renewable resources, rising to 3.75% for 2009–10, and 5% in 2010–11. A number of decisions have yet to be taken on the detailed design of the RTFO, but the feasibility studies undertaken mention the importance of an environmental assurance scheme and the sustainability of supplies (Bauen *et al.*, 2005; SDC, 2006). Reporting upon carbon saving and sustainability, both environmental and social, will be obligatory. Meanwhile, environmental groups such as WWF have called upon the EU to require certification of all biofuels consumed in member states, to demonstrate their compatibility with the sustainable development principles of the Biofuels Directive.

However, Britain is a densely populated place in world terms, and its agriculture, though productive, is relatively expensive compared to many other countries. At roughly one tonne of oil equivalent *per capita,* the UK's transport energy needs are more-or-less in line with those of other large EU economies, though less than half those of the average North American citizen (Scurlock, 2005). Can biofuels ever supply more than a few per cent of this amount?

## Ethanol and biodiesel – useful but limited potential

We should begin by reviewing the status of the most technically-proven liquid biofuels, the first-generation options that offer a pathway to a low-carbon economy which can deliver carbon savings relatively quickly.

Ethanol fuel, distilled from fermented grain or sugar crops, has been blended with petrol (gasoline) since the 1930s. Several large national programmes (Brazil, USA, several African countries) have been operating since the late 1970s, and France and Spain have emerged as the leading European producers since the late 1990s. Worldwide annual production is currently around 40 billion litres (32 million tonnes), with Brazil and the USA the two largest producers, each at about 17 billion litres per year.

Vegetable oil-based fuels have been around for decades, too. Since the 1970s, many long-term engine trials have demonstrated the feasibility of diesel/vegetable oil blends, but recent large-scale developments have been concerned with the production of biodiesel (vegetable oil methyl ester), either from recycled cooking oil, of which as much as 100,000 tonnes may be recoverable in Britain, or from virgin oils such as soybean, rapeseed or palm oil. Current world production is around 4 billion litres (3.5 million tonnes) per annum.

In fact, the history of liquid biofuels goes back to the dawn of the automobile in the late 19th and early 20th centuries. Henry Ford expected his first Model T car to be powered by grain-derived ethanol in the rural USA, and Rudolf Diesel designed his prototype engine to run on peanut oil. Such has been the power of the oil industry that it has completely taken over our perception of the internal combustion (IC) engine, although future developments are now anticipated that will optimise IC engines to match the characteristics of the new biofuels.

Both ethanol and biodiesel can be used in conventional IC engines with little modification. Blends of 5% with petroleum-derived gasoline and diesel fuel are accepted by all major motor manufacturers, and many public-sector vehicle fleets in the UK already use 5% (B5) biodiesel, since it offers a simple first step towards reducing carbon emissions from transport operations. Higher blends (B20 and B30) are now being trialled in the UK, USA and elsewhere. Volumetric blends of up to 85% ethanol with gasoline are compatible with modern flexible-fuel vehicles (FFVs), which have engine management systems that automatically adjust to a wide range of fuel blends. Millions of FFVs have been sold worldwide in the past decade, and two major manufacturers have recently introduced models to the British market.

I have discussed more widely elsewhere a range of previous assessments of Britain's liquid biofuels potential (Scurlock, 2005). In the mid-1980s, the UK Energy Technology Support Unit (ETSU) concluded that production from surplus food crops was not economically viable, based upon then current raw material prices (Marrow *et al.*, 1987).

However, even this early report determined that 5% of British petrol (gasoline) demand could be met using bioethanol from grain without disturbing existing agricultural markets. A follow-up report suggested that more biofuels could be produced from modified conventional forestry, energy crops, surplus straw and wood waste – together, enough to meet an additional 14% of contemporary gasoline demand. A later ETSU report on biodiesel (Culshaw and Butler, 1992) concluded that 6% of the UK diesel market could be replaced with biodiesel from rapeseed grown on set-aside land.

In a more contemporary review of renewable transport fuel options for the UK Department of Trade and Industry, Woods and Bauen (2003) acknowledged that existing methods for commercial production of bioethanol and biodiesel would deliver only limited short-term carbon benefits. These findings, that only a fraction of transport fuels are likely to be replaced using first-generation biofuels, are consistent with those of studies in the United States. Kheshgi *et al.* (2000) suggested that the displacement of carbon emissions from USA transport energy use was unlikely to exceed 10% of the total using domestic biofuels and current technology. Similarly, a life-cycle analysis by MacLean *et al.* (2000) found that biodiesel and ethanol have limited prospects while using food-based feedstocks. Nevertheless, the British government Energy White Paper of 2003 identified biomass-based fuels, besides renewable hydrogen used in fuel cells, as the major technological option for non-fossil transport energy (DTI, 2003).

## Beyond the RTFO

Apart from bioethanol and biodiesel, there is a diverse range of feasible routes from biomass feedstocks to transport fuels. Other possibilities include biomass gasification, followed by Fischer-Tropsch synthesis of the resulting hydrogen/carbon monoxide mixture to either gasoline or diesel substitutes, dimethyl ether (a gaseous diesel fuel made from methanol), and lastly hydrogen from biomass gasification and reforming, for consumption in either an internal combustion engine or a fuel cell (Tijmensen *et al.*, 2002; Woods and Bauen, 2003).

The term 'second generation biofuels' is generally applied to carbon-containing fuels derived from conversion of lignocellulose (the fibrous or structural component of biomass), which may be extracted from agricultural by-products, residues, wastes or energy crops. They still share most of the combustion characteristics of first-generation biofuels, and are suitable for conventional internal combustion (IC) engines.

Another term, 'next generation biofuels', has been used recently to describe other IC engine fuels made by novel conversion routes, such as microbial fermentation of sugar and starch feedstocks to bio-butanol. Energy and chemicals giants BP and DuPont are working with British Sugar to modify an ethanol production plant to produce bio-butanol by an established bacterial fermentation process. Future production is anticipated from a similar range of sugar, starch and lignocellulosic feedstocks as are used or proposed for ethanol. Butanol is claimed to offer significant advantages over ethanol for blending and transport by pipeline, due to its lower vapour pressure, its reduced susceptibility to separation in the presence of water, and its higher energy content.

Second-generation biofuels also have prospects of breaking into the technically-difficult aviation fuel market. A recent feasibility study into a range of bio-based and other alternative aviation fuels found that hydrogen, synthetic kerosene and biodiesel offered the greatest potential benefits (Saynor *et al.*, 2003). Synthetic kerosene, made from coal, is already used in aircraft departing from Johannesburg, South Africa.

The EU Biofuels Directive (proposed November 2001, adopted May 2003) required that member states should introduce targets for domestically produced liquid fuels by the end of 2004. Indicative targets were 2% of liquid fuel supply (energy content basis) by December 2005, and 5.75% by the end of 2010. Given that the voluntary target for 2010 is likely to be missed by many EU members, what policy measures may be expected in the future? The EU Strategy for Biofuels, launched in February 2006, hinted that mandatory biofuels targets are one option. A reference target of 8% by 2015 has already been proposed, following the French goal of 10% by that year. It is noteworthy that one of the three main aims of the EU strategy is to increase research into second-generation biofuels. Some preparatory work is also under way for the next stage of biofuels implementation, such as a proposed review of EU quality standards for diesel and gasoline fuels, to allow incorporation of 10% biofuel blends (the current EU diesel standard allows only 5% biodiesel).

Mandatory or voluntary policy measures similar to the RTFO and Biofuels Directive, aimed principally at first-generation biofuels, have been implemented in a wide range of countries outside Europe, including the US, Brazil, Colombia, Malaysia, Thailand and Japan. In the USA, increased market penetration for biofuels (e.g., beyond 25–30 billion litres/year) is anticipated through a gradual transition to lignocellulosic fermentation to ethanol, together with the introduction of energy-efficient electric drivetrains, etc. (Shapouri *et al.*, 2002). The period 2005–06 is now being regarded by many as a turning point for biofuel policies worldwide (REN21, 2006). Prospects are improving, and second-generation biofuels are widely anticipated in the decades between 2010 and 2030, in response to likely shortages of feedstocks for first-generation feedstocks, with hydrogen-based technologies expected to become more significant thereafter.

Some concerns remain about the degree of emissions reduction achieved with first-generation biofuels. In the medium term, second-generation fuel options such as ethanol from lignocellulosic feedstocks and synthetic diesel from gasified biomass offer better carbon savings (Woods and Bauen, 2003). Although all biomass fuels are theoretically 'low-carbon', the net energy balance (ratio of energy output to fossil energy input) varies from around 1.25 : 1 for ethanol from maize, through about 3 : 1 for biodiesel, to as much as 30 : 1 for woody energy crops like short-rotation willow. Thus, as well as offering a higher potential fuel yield per hectare (by virtue of converting the entire plant rather than just an extractable portion), second-generation biofuels offer more substantial reductions (around 90%) in 'well-to-wheels' $CO_2$ emissions compared with fossil fuels. Total greenhouse gas emissions from second-generation biofuels production using lignocellulosic feedstocks could be very low, leading to emissions reductions greater than 70% compared with fossil fuels (E4tech, 2006).

## The third generation

Third-generation biofuel production systems have been variously described as involving modified energy crop feedstocks (such as low-lignin trees or selected microalgae), modified micro-organisms or nanotechnology-based conversion systems, or (most likely) both of these. Other third-generation concepts include scaled-down bio-refineries that may be able to process a wide variety of locally-sourced feedstocks, preferably integrated with high-efficiency conversion technologies that can deliver energy services directly to the user. Utilisation of methanol or ethanol directly in fuel cell vehicles or other devices would be one such example. However, for many authorities, the term 'third-generation biofuels' is limited to hydrogen fuel produced from biomass feedstocks (crops, residues and wastes) by a variety of pathways, and utilised in either IC engines or fuel cells.

Alcohol-based fuel cells may be direct or indirect. The former utilise a separate 'reformer' stage to generate hydrogen from the methanol or ethanol fuel, and the derived hydrogen then reacts at the fuel cell electrodes. In a direct-ethanol fuel cell (DEFC) the ethanol fuel is not reformed, but fed directly to the fuel cell electrodes, where it is oxidised. Immobilised enzymes 'strip off' the hydrogen from the alcohol molecules, or nanostructured electrocatalysts achieve the same objective, all in close proximity to the electrochemical conversion reaction – so free hydrogen is barely physically present. Apart from its low toxicity and comparative safety in storage, the energy density of ethanol is many times greater than even highly compressed hydrogen, so there may be many good reasons for encouraging the development of DEFC technology.

Some authorities still expect the internal combustion (IC) engine to dominate world road transport well into the 21st century, while others point to recent advances in automotive PEM (proton exchange membrane) fuel cell stacks, which can now offer power densities of 1 kW/kg, comparable to IC engines. At some point in the future, presently predicted between about 2020 and 2030, fuel cells powered by renewable hydrogen or other third-generation biofuels are expected to make significant inroads into transport markets. Fuel cell vehicles offer possibly the best long-term potential for low-carbon transport given the extremely wide range of possible hydrogen sources. Biofuel fuel cells are expected to demonstrate significantly improved 'well-to-wheel' carbon and energy balances because of the increased conversion rate of the fuel cell compared to the internal combustion engine.

Renewably-generated hydrogen fuel is already included within the terms of the British RTFO, but this alone is not expected to be a sufficient condition for the successful introduction of third-generation biofuels in the UK (E4tech, 2006). Indeed, it has been argued that liquid biofuels should be the preferred option for providing low-carbon transport until such time as there is a surplus of renewable electricity (Eyre *et al.*, 2002). These authors suggested that too rapid an expansion of hydrogen as a transport fuel could divert renewable electricity away from displacing fossil-fired power generation, where it is more effective at reducing carbon emissions.

The 'Hypercar' concept developed by the American energy guru Amory Lovins represents a further step-change in thinking about personal transport. The Hypercar would incorporate many of the anticipated improvements in motor vehicle efficiency, as well as hydrogen fuel cells for motive power (Lovins and Cramer, 2004). Lovins recognises also the collective magnitude of the energy-conversion assets tied up in private cars, and suggests that large parts of this proposed

Hypercar fleet could be used to generate electricity for public networks when standing idle. But even this leap of the technocratic imagination cannot address the problem of infrastructural congestion (see below).

## Resources, policy measures and sustainability

The UK's biomass resource – i.e. biomass used for electricity and heat as well as for transport – is one of its largest potential sources of near-market renewable energy, second only to the wind resource. The Royal Commission on Environmental Pollution estimated that 7.5 million tonnes (Mt) of straw and 2.5 Mt of forestry timber, supplemented by a growing contribution of up to 55 Mt of energy crops, could supply an eventual total of 16 GW of heat and power generation (RCEP, 2004). The following year, the biomass task force commissioned by the British government identified 20 million tonnes (Mt) of potential feedstock with an energy content of 220 PJ, of which 8 Mt of energy crops and 4 Mt of wood were likely to be accessible (BTF, 2005). Biogas from agricultural manures and food wastes (which may also contribute towards the RTFO) is also a significant potential resource – up to 30 Mt/year dry weight – but the capital cost of required changes to vehicle and refuelling infrastructure has held it back in this regard (NSCA, 2006).

In advance of the RTFO announcement, I suggested that 5% of UK arable land, supplemented by a similar area of set-aside land, could meet a short-term goal of 2–3% of British transport fuel needs (Scurlock, 2005). More recently, the UK National Farmers Union has estimated that the RTFO requirements could be met from 375,000 ha of wheat and 840,000 ha of oilseed rape – or a net land requirement of 900,000 ha, after animal feed by-products are taken into account (NFU, 2006).

The latest research commissioned by the UK Department for Transport suggests that indigenous resources (based upon a maximum growing area of 4 million ha and an average energy yield of 125 GJ/ha) could supply up to 500 PJ of energy, about 30% of road transport energy use in 2002 (E4tech, 2006). However, the complete substitution of UK petrol and diesel supply by biofuels, around 2050, could require about two-thirds of this to be met by imports. Alternatively, the total UK renewable resource base (i.e., biomass supplemented by other renewables) could contribute as much as 40–80% of future transport needs (Woods and Bauen, 2003).

Although oil still accounts for more than 96% of transport energy worldwide, biofuel production has doubled over the past five years and is poised for significant further growth. In 2005, world biofuel production surpassed 670,000 barrels per day (about 39 billion litres per annum), equivalent to around 1% of the global transport fuel market (Worldwatch, 2006). By 2030, the International Energy Agency estimates that up to 7% of road transport worldwide could be powered by biofuels, with Brazil, the USA and the EU likely to remain the largest producers and consumers (IEA, 2006b). Brazil presently leads the world in biofuel market share, at more than 25% of total road fuel use, and plans to increase fuel ethanol production by at least 50% in the coming years. Proposals include an expansion of exports to Europe and the USA (about 15% of production – 2.6 billion litres – was exported in 2005). In response to concerns about the sustainability of its strategy, the Brazilian government points to the modest proportion of land area under sugar cane (0.6% of total land) and the concentration of sugar enlargement in areas of degraded pasture, far from the environmentally-sensitive tropical forests (Embassy of Brazil, 2006). A recent study suggests that such international trade in transport biofuels may be justifiable on environmental as well as economic grounds, since at least some Brazilian ethanol supply chains already meet Dutch standards of sustainability (Smeets *et al.*, 2006).

Other environmental concerns have been raised about the sustainability of biodiesel feedstock supply. According to Friends of the Earth (FoE), food industry demands for palm oil already threaten the forests of South East Asia, and a projected steep rise in additional use for energy will only create further conflict between palm oil companies, local communities and wildlife such as the orang-utan. FoE has requested that governments in Malaysia and Indonesia demonstrate that carbon-storing rainforest is not being cleared to make way for oil palm plantations. The industry response has been rapid, with the establishment in 2004 of the Roundtable on Sustainable Palm Oil, a body that has since developed sustainability criteria and verification protocols for its membership, which includes a worldwide range of growers, processors, manufacturers and retailers.

Indeed, it may be argued that enlargement of tropical cash crops at the expense of tropical forests and other sensitive habitats has been continuing since the 1960s, largely as a result of expansion in the world food industry. Thus the damage has already been done – and the addition of biofuel crops to a wide diversity of fodder and food crops (plus other non-food commodities like tobacco, cocaine, etc.) may be an opportunity to improve the situation through demands for carbon accounting and sustainability criteria. Similarly, domestic demand for certified low-carbon bioenergy crops in Europe or the USA may allow the partial recovery of some landscapes from intensive food cropping. It is only logical that environmental policies to address carbon emissions to the atmosphere must necessarily be accompanied by checks and balances to avoid other environmental disbenefits.

It is important for key stakeholders (farmers, petrochemicals companies, fuel distributors, etc.) to engage with the environmental movement to garner their support for biofuels and the associated agricultural supply chain. Equally, it is pragmatic for environmentalists to find common ground with the motor industry, the fossil fuel suppliers and the agricultural sector in order to deliver sustainable development of the transport sector. Concerned about the scale of future biofuel plantations and tacit support for continued expansion of private transport, some in the environmental movement are critical of others for forming an unlikely alliance with farmers and industry to pave the way towards more resource-efficient second-generation biofuels. But this is a logical response to the perceived risk that the transport fuels sector, already prone to technological lock-in, may stick with first-generation technology in the absence of carefully constructed regulatory incentives for investment in the next generation of production methods. Growers, technology companies and policy advisers alike are already concerned that without an appropriate carbon assurance scheme the current round of directives and incentives may lock second-generation technologies out of the market. Since they are driven by 'green' imperatives, it is also clear that the emerging regulatory frameworks should emphasise and support other environmental measures alongside the production of non-food crops in Europe, such as eco-friendly land management techniques (restoration of hedgerows, wider field margins, etc.).

In the future, the availability of biomass feedstocks for transport energy is likely to be limited by competition from the heat and power generation sectors, at least at the national level. Meanwhile, on the international stage, there will be competing demands for food, fibre (everything from timber to textiles) and a range of different biomass energy feedstocks. The dynamism of the transport market may indeed help to create the opportunities for development and better management of biomass raw materials, while setting standards for sustainability – but this is far from a foregone conclusion. There is still a danger that, with hindsight, the shift towards biofuels may be regarded as nothing more than a land-grab to satisfy the thirst of the world's growing population of oversized private cars, SUVs and pickups – unless more is done about energy efficiency (see discussion below).

Other forecast threats to future biofuel availability include its sensitivity to disruption by drought or pests, with climate change throwing in a further measure of uncertainty – although it is true that world food supplies have always been subject to these same pressures. Again, perhaps the pre-eminence of energy security may help to set new standards for food security worldwide. As humanity engages with the early signs that the Earth has real, tangible limits to growth, we may at last learn the art of wholesale planetary management.

Now that it looks like many countries are on track to meet their initial biofuel objectives, the world can afford to speculate about what happens next – but until we get a little closer to those 2010 targets, we don't really know how quickly the nascent biofuels industry will respond to market stimulation measures. The real challenge lies beyond that – replacing 50% or more of petroleum feedstock. At that level, there will be a need for integrated policies covering energy, environment and food production. Recent scenarios that predict the ultimate techno-economic potential of a range of biomass energy feedstocks in the context of world food and fibre production suggest as much as 400 EJ could be available by 2050 – about 85% of current worldwide primary energy supply. Such estimates assume a shift towards perennial energy crops, technological improvements to crop yields and greenhouse gas balances, and further expansion of international trade in biofuels (Junginer *et al.*, 2006). This could bring significant benefits to developing countries, in terms of better prices for low-carbon energy commodities, but substantial government intervention may be required to avoid undue competition between production of solid and liquid biofuels, while maintaining security of food supply, biodiversity and optimum use of land for carbon management.

Some of these advantages and drawbacks are summarised in Box 3.1.

## Efficiency, congestion and personal freedom

It is clear that the Earth cannot sustain an expansion of the transport services typically demanded by advanced economies to the world at large without a dramatic increase in end-use efficiency. Energy efficiency is one of the major technological drivers of sustainable development worldwide, but the growing demand for personal mobility and leisure travel continues to outstrip technical and social adjustments in most parts of the world (UNDP, 2001). Perhaps it is rather premature for the UNDP World Energy Assessment to assert that 'by combining

*Box 3.1* **The case for biofuels – arguments for and against**

**For**

- reduce $CO_2$ emissions by 55–80%, depending upon methods of production and utilisation
- increase diversity and security of energy supply
- opportunities for diversification and added value in agriculture
- in general, positive impact upon local air quality
- transitional step towards full sustainability of transport provision

**Against**

- vulnerability of feedstocks to climate, pests, etc.
- competition for land use with food crops and other biomass feedstocks
- difficulty in assessing $CO_2$ saving and sustainability from some sources
- potential impact on land use and biodiversity in developing countries

new fuels....with improved modes of transportation and vehicle performance, it appears possible to meet sustainability criteria' (UNDP, 2004).

Furthermore, the congestion of road and rail networks, and even airport facilities, are a symptom of the finite capacity of the infrastructure to sustain further development. Business travel planning, local government planning of transport provision and public transport incentives are all measures that may help to constrain demand, but they are often perceived as marginal, with only incremental impact upon rates of transport growth. Demand-side management is an art yet to be applied effectively to the transport sector, despite efforts to develop computerised car-pooling, as well as other measures that encourage consumers to switch from private to public transport or alternative low-carbon travel modes such as cycling and walking. A combination of both improved technology and switching of transport modes may be required to produce significant reductions in carbon emissions, since neither has sufficient potential to deliver sustainable transport alone (Potter, 2006). Besides, national and local climate change policy may be at odds with planning for economic development, as is evident in the growth of regional airports in the UK. A 2005 report by the Tyndall Centre for Climate Change Research concluded that all other sectors of the economy would have to reduce their carbon emissions to zero, if the aviation industry were to be allowed to grow within the context of climate change targets (Anderson *et al.*, 2005).

In the future, we may even have to recognise that there are limits to the level of transport services that the 'world citizen' can reasonably demand. This raises some rather provocative questions. Perhaps future measures to constrain transport demand could impact upon our perception of civil liberty and human rights? Should the ability to travel to other parts of one's own country, or indeed to other continents, whether for tourism or to seek new work and life opportunities, be regarded as a right or a privilege? It is not only technological and structural changes that we face in the future of transport services, but social and political change as well.

## References

Anderson, K., Shackley, S., Mander, S. and Bows, A. (2005) Decarbonising the UK: energy for a climate conscious future. Technical Report 33, Tyndall Centre for Climate Change Research, Norwich. 89 pp.

ANRE (2006) Energy Supply in Japan Today. Agency for Natural Resources and Energy, Government of Japan, Tokyo. http://www.enecho.meti.go.jp/english/energy/japan/supply.html (accessed 1-Oct-06).

Bauen, A., Howes, J., Chase, A., Tipper, R. Inkinen, A., Lovell, J. and Woods, J. (2005) Feasibility study on certification for a Renewable Transport Fuel Obligation, final report, June 2005. http://www.dft.gov.uk/stellent/groups/dft_roads/documents/divisionhomepage/610328.hcsp.

BTF (2005) Final report to Government. Biomass Task Force, York. 80 pp.

Culshaw, F. and Butler, C. (1992) *A Review of the Potential of Biodiesel as a Transport Fuel*, ETSU R-71, The Stationary Office, Norwich.

DEFRA (2006) UK Emissions of Greenhouse Gases. Department for Environment, Food and Rural Affairs, London. http://www.defra.gov.uk/environment/statistics/globatmos/gagccukem.htm (accessed 1-Oct-06).

DTI (2003) Our Energy Future: creating a low-carbon economy, Department of Trade and Industry Energy White Paper, February 2003, Cm 5761, The Stationary Office, Norwich.

DTI (2006) Department of Trade and Industry *Digest of UK Energy Statistics*, 2005, The Stationery Office, Norwich, 2006.

E4tech (2006) UK Carbon Reduction Potential from Technologies in the Transport Sector. Report to UK Department of Transport and Energy Review team, May 2006. E4tech, Imperial College, London. 54 pp.

ECI (2006) Predict and Decide: aviation, climate change and UK policy. Environmental Change Institute, University of Oxford. 122 pp.

EIA (2006) Petroleum Products Information Sheet. Energy Information Administration, US Department of Energy, Washington DC http://www.eia.doe.gov/neic/infosheets/petroleumproducts.htm (accessed 1-Oct-06).

Embassy of Brazil (2006) Clean Energy: the Brazilian ethanol experience. Embassy of Brazil, London. 72 pp.

Eyre, N., Fergusson, M. and Mills, R. (2002) *Fuelling Road Transport: implications for energy policy,* Energy Saving Trust, London / Institute for European Environmental Policy, London / National Society for Clean Air, Brighton.

IEA (2003a) Energy Balances of Non-OECD Countries 2000–2001. International Energy Agency, Paris.

IEA (2003b) Energy Balances of OECD Countries, 2000–2001. International Energy Agency, Paris.

IEA (2006a) Energy Statistics and Energy Balances, by country. International Energy Agency, Paris. http://www.iea.org/Textbase/stats/index.asp (accessed 1-Oct-06).

IEA (2006b) World Energy Outlook 2006. International Energy Agency, Paris. 600 pp.

Junginger, M., Faaij, A., Rosillo-Calle, F. and Woods, J. (2006) A growing role: opportunities, challenges and pitfalls of the biomass trade. *Renewable Energy World* 9(5), Sept–Oct 2006.

Kheshgi, H.S., Prince, R.C. and Marland, G. (2000) The potential of biomass fuels in the context of global climate change: focus on transportation fuels, *Annual Review of Energy and the Environment,* 25, 199–244.

Lovins, A.B. and Cramer, D.R. (2004) Hypercars, hydrogen, and the automotive transition. *International Journal of Vehicle Design* 35, 50–85.

MacLean, H.L., Lave, L.B., Lankey, R. and Joshi, S. (2000) A life-cycle comparison of alternative automobile fuels, *Journal of the Air and Waste Management Association* 50, 1769–79.

Marrow, J.E., Coombs, J. and Lees, E.W. (1987) *An Assessment of Bio-ethanol as a Transport Fuel in the UK,* ETSU R-44, The Stationary Office, Norwich.

MOSPI (2006) Energy Statistics 2004–2005. Ministry of Statistics and Programme Implementation, Government of India, New Delhi. http://mospi.nic.in (accessed 1-Oct-06).

NFU (2006) UK biofuels – land required to meet RTFO 2010. NFU press release, 10-Aug-06, National Farmers Union, Stoneleigh, Warwickshire. http://www.nfuonline.com/x9761.xml.

NSCA (2006) Biogas as a Transport Fuel. National Society for Clean Air and Environmental Protection, Brighton. 52 pp.

Potter, S. (2006) Travelling Light: the roles of behavioural change and technical innovation in achieving sustainable transport. Keynote paper for the 2006 Shell Eco-marathon. The Open University, Milton Keynes.

RCEP (2004) Biomass as a renewable energy source. Special Report, Royal Commission on Environmental Pollution, London. 92 pp.

REN21 (2006) REN21 2006 Global Status Report. Renewable Energy Policy Network for the 21st Century, Paris. 35 pp.

Saynor, B., Bauen, A. and Leach, M. (2003) The Potential for Renewable Energy Sources in Aviation, Report to Department of Trade and Industry. ICCEPT, Imperial College, London. 78 pp.

Scurlock, J.M.O. (2005) Biofuels for transport in the UK: what is feasible? *Energy and Environment* 16, 273–82.

SDC (2006) Response to the Department for Transport on Biofuels and the Renewable Transport Fuels Obligation, June 2006. Sustainable Development Commission, London. 24 pp.

Shapouri, H., Duffield, J.A. and Wang, M.Q. (2002) *The Energy Balance of Corn Ethanol: An Update*, U.S. Department of Agriculture, Economic Research Service, Agricultural Economic Report No. 813.

Smeets, E., Junginger, M., Faaij, A., Walter, A. and Dolzan, P. (2006) Sustainability of Brazilian bio-ethanol. Copernicus Institute, University of Utrecht, The Netherlands / State University of Campinas, Brazil. 136 pp.

Tijmensen, M.J.A., Faaij, A.P.C., Hamelincka, C.N. and van Hardeveldb, M.R.M. (2002) Exploration of the possibilities for production of Fischer Tropsch liquids and power via biomass gasification. *Biomass and Bioenergy* 23, 129–52.

Woods, J. and Bauen, A. (2003) *Technology Status Review and Carbon Abatement Potential of Renewable Transport Fuels in the UK*, Report B/U2/00785/RP, Department of Trade and Industry, London.

Worldwatch (2006) Biofuels for Transportation: Global Potential and Implications for Sustainable Agriculture and Energy in the 21st Century. Worldwatch Institute, Washington DC.

UNDP (2001) World Energy Assessment: Energy and the Challenge of Sustainability. Sustainable Energy and Development Division, United Nations Development Programme, New York. 516 pp.

UNDP (2004) World Energy Assessment: Overview – 2004 update. Sustainable Energy and Development Division, United Nations Development Programme, New York. 88 pp.

# 4

# Sea Power – How We Can Tap Wave and Tidal Power

*David Elliott*

Renewable energy can be used both in rural and urban areas. Given proper attention to the efficient use of energy, most rural areas can probably meet many of their needs from solar collectors, solar photovoltaics, biomass and biofuels, and, depending on their location, local wind turbines and micro hydro plants, with the national grid system helping to even out local variations in supplies and demand. Given proper attention to the efficient use of energy, most cities can also produce much of their energy from roof top solar heating and PV electricity systems, and from advanced waste-into-energy systems like pyrolysis and gasification – after all, waste is one thing in which cities are self sufficient (Elliott and Taylor, 2000).

However there will still be a need for some larger power inputs in the case of urban areas, and also possibly for some rural areas. At present in the UK we rely on large fossil and nuclear fuelled power stations, mostly located in rural areas. Renewable energy options like large wind farms and energy crop plantations can help. They too will have to be in rural areas, which is where the main resource is found. However there are limits to how much land can be used for that purpose. Cities inevitably rely on rural areas for many of their resources, most obviously food and water. But some rural residents have already begun to complain about the local environmental and amenity losses inflicted by large wind projects. Of course, these local impacts are small compared with the global impacts associated with the use of fossil fuels, but even so, there will inevitably be limits to how much land we can use for energy production – since there are other competing uses, not least food production, but also residential space and leisure space.

Fortunately, the UK is blessed with a range of offshore renewable energy options which can alleviate this problem. The **offshore wind** resource is very large – it could supply at least 20% of UK electricity needs and possibly much more. The UK has identified 18 potential sites for development around the coast and has designated three special strategic zones for detailed assessment. So far, three wind farm proposals have gone ahead and more are in the offing. Similar projects are underway elsewhere. Altogether, by 2006, there was nearly 700 MW of offshore wind generation capacity installed in the EU, with the largest project so far being the new 270 million Euro Horns Reef project, stretching between 14 and 20 km off the coast of Denmark. Many more large projects are underway or planned, including projects in the US and China.

Wave energy is an equally large, but, so far, less developed option – the global potential has been put at 2,000–4,000 TWh pa, with, given its Atlantic setting, the UK having one of the best resources. In the 1970s there was a flurry of activity in the UK, but the wave programme was halted in the 1980s following some adverse, and disputed, assessments of likely costs. But, now, after a 25 year delay, new projects are coming forward. Similarly for tidal power. That too was backed in the 1970s, with the focus on large tidal barrages, like the proposed 8.6 gigawatt Barrage across the Severn estuary, but abandoned in the early 1990s as likely to be too expensive and environmentally invasive (Elliott, 2004). But now the emphasis is on the new concept of tidal current turbines, operating on the tidal flows, and also possibly offshore tidal lagoons, rather than barrages.

This chapter looks at how the wave and tidal dream has fared from the early days to the present – and at what might emerge in the future. The story is one of a series of technologies that have come from the margins, and overcome major institutional and well as technical obstacles.

## Picking winners

Following the 1974 oil crisis, the UK government launched a renewable energy assessment and support programme, with the main focus being wave energy and tidal barrages. There were also geothermal energy and solar energy programmes. At that time windpower was seen as at best a marginal possibility. In 1973 the UK Department of Energy had concluded, in Energy Paper 21, that *'although aerogenerators might be considered economic on certain hill sites... a clear economic case cannot be made for a programme large enough to make a significant contribution to the nations' energy supply.'* Some innovative wind projects nevertheless emerged, but were not followed up.

Thirty years on however, wind power has become a major success around the world – it is currently one of the fastest growing new energy technology, with, by 2006, around 60,000 megawatts of generating capacity installed, but with the UK struggling to catch up, and having to import wind turbines from Denmark and Germany.

Unfortunately, the UK also seemed to risk losing its initial lead in the wave power field. The UK wave energy programme was initiated in the mid-1970s under a Labour administration and led to some pioneering designs, including Stephen Salter's Duck system, some of which were tested as scale prototypes in open water. However, following the election of a Conservative government committed to cutting public expenditure, an assessment in 1982 by ACORD, the governments Advisory Committee of Research and Development, concluded that further work on deep sea wave energy should be halted, on the basis of some high estimates of likely generation costs (estimates of 20 p/kWh and even 50 p/kWh were mentioned). This assessment was strongly challenged at the time. Some critics claimed there was a pro-nuclear bias involved, and that, at the very least, the technology had been assessed at too early a stage in its development. The all-party House of Commons Select Committee on Energy commented in 1984 that the suspicion was that wave energy *'was effectively withdrawn before the race began.'* (Select Committee, 1984).

Certainly there was room for disagreement amongst the experts: they faced the problem of trying to cost a range of very novel systems. As a spokesman for the Department of Energy was to tell a session of the same Select Committee some years later, there *was 'definitely scope for different judgements at the early stage of the development of a device.'* (Select Committee, 1992). Nevertheless, in 1992, a new review of what remained of the wave energy programme decided that it too should be wound up.

In 1994, the governments tidal barrage programme was also wound up, following a long running programme of assessment of the Severn, and other smaller potential sites, on the grounds that barrages would be environmentally intrusive and also hard to finance given their large capital costs. Certainly the electricity privatisation programme and competitive market ethos of the 1990s, made private sector investment as unlikely as public sector support. Apart from a small independently funded project looking at the idea of tidal current turbines, the tidal option seemed to be off the agenda.

It was not until after a change of government, in 1997, that views began to change. Although given the funding cutbacks, not a lot of new work had been done on wave or tidal energy in the UK in the meantime, the political climate had clearly changed, in part because of growing concerns about climate change. A series of reassessments of wave energy and tidal energy were carried out, initially as part of the UK Technology Foresight programme, culminating in March 2001, with an admission by the Department of Trade and Industry that the decision in 1992 to abandon wave energy was wrong:

> The decision was taken in the light of the best independent advice available. With the benefit of hindsight, that decision to end the programme was clearly a mistake (DTI view quoted in Ross, 2002).

Wave power has now been resuscitated as a option, and tidal current energy is also being pursued with more enthusiasm, and although it is still too early to say whether these new sources will prove as successful as hoped, their combined energy potential is large, and the initial decision to downplay these options does seem to have been premature. Certainly that was the view of the House of Commons Select Committee on Science and Technology, which in 2001 commented bitterly that *'given the UK's abundant natural wave and tidal resource, it is extremely regrettable and surprising that the development of wave and tidal energy technologies has received so little support from the Government'*. (Select Committee, 2001)

## The UK's new wave and tidal programmes

The new programmes have adopted a very different approach from that in the 1970s. Rather that large-scale deep-sea wave energy systems and giant tidal barrages, the focus has been on small inshore and onshore wave devices and small modular tidal turbines sited in tidal flows.

First in the water was Wavegen, who, following a smaller prototype unit, installed a 500 kW 'oscillating water column' ('OWC') shore-line wave device, the Limpet, on the coast of the Isle of Islay. Wavegen is now planing a 3 MW version for location at Siadar on Lewis in the Outer Hebrides, in a joint project with npower renewables. Wavegens' earlier work on inshore devices has also been continued, although the failure of the Osprey sea-bed mounted prototype in 1995 (it was damaged by a storm before it could be secured to the sea bed), has led to more interest being shown in floating devices, which are seen as better able to withstand rough seas. For example, Ocean Power Delivery (OPD) have developed a floating 'sea snake' articulated cylinder wave energy system – the Pelamis. Eventually the idea is to have a 40 unit 30 MW wave farm – Portugal has already commissioned a system with initially three 750 kW units installed 5 km off the coast. OPD have also been chosen as one of three candidates for using the proposed Wave Hub off the north coast of Cornwall. They plan to trial up to 10 Pelamis devices there. The Hub is in effect a sea-bed mounted power socket, with connections to land, into which developers will be able to feed power, rather than each having to make their own sea to shore connections. The other wave projects to be linked to the Hub involve the US company OPT, who will install their PowerBuoy system, and the Norwegian company Fred Olsen Ltd, who have developed a floating wave buoy/raft system.

As can be seen the UK, once the clear leader, has been joined by developers from other countries in the wave energy field, although some ideas developed elsewhere have also been taken over by UK developers. For example, in 2004 a Scottish company, AWC Ocean, took over development of the Archimedes Wave Swing device, which had originated in the Netherlands. In addition, £5 million in EU Objective 1 funding has been made available for a 7 MW demonstrator project based on the Danish Wave dragon design, to be installed off the coast of Wales near Milford Haven. If this is successful, there are plans for installing more, in a 70 MW wave farm.

New ideas also continue to emerge, like the Oyster sea bed mounted device being developed by Queens University Belfast, one of the other wave power pioneers.

However, to some extent, with some wave energy projects moving towards the commercial deployment stage (as with the OPD Pelamis project in Portugal), the innovative emphasis has moved on to **tidal power** and the development of tidal current turbines of various types. Following on from the success of the 240 MW tidal barrage on the Rance estuary in France, in the 1970s the main way in which it was thought possible to harness tidal power was *via* large barrages across tidal estuaries, trapping the high tide and letting it out through turbine generators as with hydro power plants. If all possible sites could be used, about 15% of UK electricity could perhaps be supplied in this way. However barrages would be expensive and could be environmentally invasive. By contrasts tidal current turbines use the horizontal tidal flows rather than the vertical rise and fall, and that means there is no need for large, expensive and invasive barrages across estuaries.

The UK has been at the forefront of the development of this idea, due mainly to the pioneering work of IT Power. In 1994, IT Power built and tested a small two bladed 10 kW prototype, with support from

Scottish Nuclear and NEL, in the Corran Narrows in Loch Linne, near Fort William in Scotland. It involved tethering the device between a floating pontoon and an anchor on the sea bed.

Subsequently IT Power set up Marine Current Turbines Ltd (MCT) to develop a new version of their 'Seaflow' turbine design, mounted directly on the sea bed. In 2002, they began testing a 300 kW device off the coast of N. Devon, at Lynmouth, with support from the DTI and the EU. The technology is based on the use of monopiles – single steel tubes which can be set in holes drilled in the sea bed, and then coupled with two or three bladed propeller units. Peak spring tidal currents of 4–5 knots (2 to 2.5 m/s) are seen as necessary, with depth of between 20 and 35 m, for economic exploitation, and the company has identified scores of potentially suitable sites around the UK coast, and elsewhere in the world.

The next stage is a larger commercial model of 1.2 MW, known as the 'SeaGen', for installation, subject to environmental assessment, in Northern Ireland's Strangford Lough, with funding support from the DTI. Following that, MCT are looking for sites for 10 MW 'tidal current farm' arrays, one possibly off Lynmouth and several possibly off the Welsh coast – sites around Anglesey for a 7 unit tidal farm are being studied, with £700,000 in grant support from the Welsh Assembly Government's Objective 1 programme.

Overall MCT have estimated that Wales could generate around 8% of it electricity from tidal current turbines, and in 2006 they told the Welsh Affairs Select Committee that they were looking to install 100 MW 'within five to eight years'. They added that 'in fact there is 1 gigawatt potentially available' (Welsh Affairs Committee 2006).

MCT are not the only developers in the field. In parallel, Swanturbine Ltd, based at Swansea University, are developing a similar design to MCTs's, but with a variable speed rotor to avoid the need for gearing, thus increasing energy capture efficiency and reducing maintenance costs and noise. In addition they have developed a novel seabed mounting arrangement to reduce installation costs. They have tested a 1 metre diameter prototype in the river Towy and are developing a 350 kW demonstration version.

Another contender for playing a major role is SMD Hydrovisions' twin rotor Tidel device. It is a tethered system, which, like the early IT Power rotor mentioned above, can swing round to face the changing tidal flows. A 500 kW prototype is being tested with support from the DTI. Several other propeller-type devices are also under development, including an array of rotors on a sea bed mounted frame being developed by Tidal Hydraulic Generators Ltd, which is based in Pembokeshire.

There has also been a flurry of innovative developments of the basic propeller/rotor concept, like Lunar energy's ducted-rotor device being tested on the Orkneys – the ducting shroud is designed to accelerate the flow of the tidal flow it captures. Open Hydro have developed and tested a novel Open Centre Turbine, which has rotor blades running in an outer ring. A version of it is to be installed off Alderney. In addition there is the SST (Semi-Submersible Turbine) concept being developed by the Tidal Stream consortium. This has an array of rotors mounted on a hinged gravity base to allow for easier maintenance access. Meanwhile, Orkney-based Scotrenewables is developing a novel free-floating, rotor-based, tidal current energy converter. In 2006 it was awarded the inaugural Springboard Award, set up by Shell to support commercially viable and innovative projects by small companies that could lead to greenhouse gas reductions. A three-year development programme supported by the Carbon Trust is expected to lead to a full-scale prototype tidal turbine demonstrator in 2008. Another innovative development in Scotland is the Tidal Delay system. This involves using electricity from a tidal turbine to produce heat to top up a graphite-based thermal store. This in turn can produce steam continuously to drive a generator, thus compensating for the cyclic nature of the tides. There are plans to test it in the Firth of Forth.

Not all developers have adopted propeller-type systems. One interesting idea was developed by the Northumbria-based marine engineering company, the Engineering Business – the Stingray, a hydroplane system with a series of fully submerged parallel wing-like fins which move up and down harvesting energy from the tidal flows. A 185 tonne 150 kW prototype, with 15 m planes, was tested off the Shetland Islands in 2002. Subsequently, Pulse Tidal has developed what is claimed to be a high-efficiency hydrofoil-type device for testing in the Humber estuary, with DTI backing. Another novel idea is the Rochester Venturi (RV) invented by Geoff Rochester at Imperial College, London. The device extracts energy from the tidal flow by using the Venturi effect in a pipe to drive a turbine connected *via* a secondary pipe.

Finally, wave energy pioneer Prof. Stephen Salter has also developed an idea for a tidal current device. Dubbed the Polo, the new device has series of vertical-axis blades mounted in a giant ring-shaped carrousel-like structure which floats on the surface, with the blades reaching into the water below and able to rotate on roller bearings in the tethered ring.

Interestingly, at the hearings on Wave and Tidal power held in March 2001 by the Select Committee on Science and Technology, Salter said that he saw tidal current systems as likely to develop more rapidly than wave technology, despite the formers earlier start. *'I think the uncertainties about tidal streams are lower, I think they can take a lot of technology from wind, and I think they are a more predictable environment; so I would expect that would reach commercial viability sooner than wave energy. The problem is that it is not such a large resource, and we can use all of it and still want more, whereas wave energy is such a big resource that it is worth going for, even if it looks hard to start with.'*

As Salter noted, the UK's tidal current resource is smaller than the wave resource. This is because tidal flows of reasonable velocity only occur in a limited number of near-shore geographically defined locations, where landmasses form constrictions to tidal flows, whereas waves are created by the action of wind moving over the open sea and this energy can be extracted in a wide variety of locations.

The energy review by the UK Cabinet Office's Performance and Innovation Unit in 2003 put the total UK *theoretical* wave and tidal current resource at 700 TWh p.a. That is nearly twice current UK electricity consumption. However not all of this could be extracted in practice. In 2004, the DTI/Carbon Trusts' Renewables Innovation Review, put the UK's total *practical* wave resource at around 70 TWh of energy per annum, about 20% of UK electricity requirements, while the practical tidal current resource was put at around 31 Twh p.a., about 10% of UK electricity requirements. Subsequent studies by the Carbon Trust, taking economic constraints into account, refined these estimates, with the offshore wave figure being put at 50 TWh, plus 27.8 TWh for near shore and 0.2 TWh for shoreline wave, and the tidal figure at 18 TWh p.a. Overall the Carbon Trust claimed that the UK might ultimately expect to obtain up to around 20% of its electricity from wave and tidal currents (Carbon Trust, 2006).

Whether this is the ultimate limit, given for example the very large wave resource available if devices can be located in the deep sea areas off the UK coast, and how rapidly the resource could be exploited, is unclear. Certainly so far the pace of development has been relatively slow, with only a few small wave and tidal projects in place, totalling, according to the 2006 Energy Review, around 7 MW, but there are signs that it could pick up. For example, the renewable energy route map produced by Scottish Renewables in 2006 claimed that industry was ready to deliver 160 MW of wave and tidal projects by the end of 2010, with presumably much more to follow thereafter in the UK and elsewhere. Indeed, long range OXERA modelling for the DTI has suggested that, on the basis of an enhanced marine emphasis in order to obtain a 20% contribution from renewables, there could be 5,000 MW of wave and tidal power capacity installed by 2025 (Swanturbines, 2006).

## Developments around the world

The UK has some of the worlds' best sites for both wave and tidal currents, but several other European countries, and others around the world, also have reasonable wave and tidal resources. The DTI/Carbon Trust put the EU's total practical wave resource at 250 TWh and the tidal resource at 48 TWh, while the total world wave resource was estimated to be 2,000–4,000 TWh, and that for tidal currents around 800 TWh (DTI/Carbon Trust, 2004).

It is not surprising then that many other countries have also been exploring these options, with export potential being an obvious attraction. For example the DTI/Carbon Trust report suggested that by 2050 there might be over 200 GW of wave and around 20 GW of tidal current generating capacity in place world wide. A report in 2005 by Douglas-Westwood consultants suggested that the EU might expect a £600 billion global market for wave and tidal technology. It put the total global tidal market as being in the range £155–444 billion and that for wave as £450–1,175 billion (ReFocus, 2005).

In the **wave energy** field, the Danes developed a Wave Dragon device (subsequently taken over by a UK company), which involves two wave reflectors focusing waves up a ramp into a reservoir, with the head of water then being used to drive turbines. The Danes are also progressing with the novel Waveplane, which funnels waves of different height into a series of channels, creating a vortex which drives a turbine. In addition, as noted earlier, the Dutch developed a novel Archimedes Wave Swing device. This consists of a linked series of submerged air chambers which rise and fall in sequence with the wave pattern, and pump air through a turbine to generate power (Smith, 2005). Test have been carried out on a prototype in Portugal, and as noted above, in 2004 a Scottish company, AWC Ocean, took over development of this device.

In Australia, a 300 kW prototype of Energetechs novel shore-based wave reflector device is being tested at Port Kimbla, 100 km south of Sydney. It is an Oscillating Water Column (OWC) device, with a variable pitch turbine. BC Hydro in Canada, have decided to use a version of this device for a 100 MW wave project in Vancouver, and the US Aqua Energy Group is planning a 1 MW project near Neah Bay in Washington State. The US company OPT has also developed a buoy unit which it is deploying off the coast of Spain. In parallel, Japan has been working in the field for many years and has installed many OWC units on breakwaters. It also has a floating wave energy test bed – the Mighty Whale (Smith, 2005).

In the **tidal field**, a propeller type device has been developed by a consortium of Norwegian companies including Statoil and ABB. Based on a system similar to that installed by Marine Current Power in Devon, a windmill-like turbine is being tested on the seabed near Kvalsund at the Arctic tip of Norway.

There have also been reports of an experimental floating tidal power being developed by Hydro WGC in Russia, with a 1,500 kW turbine system having been launched at the Sevmash Plant in the city of Severodvinsk, in the Arkhangelsk region and then towed for testing near the existing Kislogubskaya Tidal Power Plant in the Barents Sea (MPS, 2006).

Meanwhile, in somewhat warmer waters, the US Current to Current Corporation is installing a novel 10 MW ducted-rotor tidal current turbine system in Bermuda.

There are also proposals from Neptune Power for an array of 1 MW floating sub-sea turbines in the tidal currents of the Cook Strait between New Zealand's North and South Islands, with ultimately up to 7,000 turbines being anchored to the sea floor 40 m below the surface. In addition, there is a proposal for installing a 200 MW array of tidal turbines just north of Auckland in the mouth of Kaipara Harbour. And in the US, Verdant Power is developing a project with six 35 kW tidal turbines for installation in New York's East River, with the UN Headquarters being one possible customer. Verdant power also has some larger projects underway elsewhere, including a 15 MW project in Canada and a 25 MW project in the Amazon Basin (MPS, 2006).

On a somewhat larger scale, the Canadian Company Blue Energy has developed a 'tidal fence' concept, in which H-shaped vertical axis turbines are mounted in a modular framework structure. They see this as being suited to installation in causeways between islands, and have developed ambitious plans for a four-kilometre long tidal fence between the islands of Samar and Dalupiri in the Philippines, with a total estimated generating capacity of 2,200 MW at peak tidal flow (1,100 MW average).

On an even larger scale, Dr Alexander Gorlov, Professor of Mechanical Engineering at Northeastern University in Boston, has been developing ideas for a novel vertical axis turbine device for use with ocean currents further out to sea, including the Gulf stream. Strictly this is not tidal power – the Gulf stream is not the result of lunar gravitational forces, but is part of the conveyor system of planetary ocean current, driven ultimately by solar heat. But there is certainly a lot of power there. Gorlov notes that *'the total power of the kinetic energy of the Gulf Stream near Florida is the equivalent to approximately 65,000 MW.'*

Finally, although conventional tidal barrages are no longer seen as attractive options by most analysts, it has been suggested that building bounded reservoirs out to sea in shallow water, rather than dams across estuaries, could be cheaper and less invasive. The US company Tidal Electric has been following up this offshore 'tidal lagoon' idea. It has plans for projects around the world, including some off the Welsh coast, with the most advanced being a proposal for 60 MW lagoon scheme in Swansea Bay. There has also been interest in this idea in China, with agreement being reached in 2004 to explore the feasibility of a 300 MW project near the mouth of the Yalu river.

## The future

It has been a long hard struggle to get wave and tidal power recognised, and even now the potential is often down-played. For example, in a review of UK renewable energy options, carried out between 1997–2000, the Department of Trade and Industry in effect relegated tidal current technology to the 'very long term' category. Despite there being more enthusiasm from Ministers these days for wave and tidal power, neither source is seen as making a significant contribution to meeting UK energy requirements before 2010, or even 2020. A study by the Governments Energy Technology Support Unit in 1999 did suggest that by 2025 the UK might expect to have 3,700 MW of wave generation capacity in place, but it estimated that there would only be 250 MW of tidal current capacity (ETSU, 1999). And in its report on UK energy policy and climate change published in 2000, the Royal Commission on Environmental Protection suggested that even by 2050 the UK might only have around 500 MW of tidal stream generation capacity in place, although wave power did better with 7,500 MW installed (RCEP, 2000). Overall, the DTI/Carbon Trust Renewables Innovation Review in 2004 suggested that wave and tidal energy might provide 12–14% of UK electricity by 2050. The most recent study, by the

Carbon Trust, suggested that by 2020 the UK might only expect to obtain around 3% of its electricity from wave and tidal current systems and, as noted earlier, only 20% ultimately (Carbon Trust, 2006).

By contrast, the enthusiasts see it very differently. During a House of Commons debate on Renewable Energy in April 2001, Dr. Desmond Turner (Brighton, Kemptown) strongly backed wave and tidal power. He was a member of the Science and Technology Committee which had been looking at these options. He claimed that *'If wave or tide is fully exploited, there is the potential to produce sufficient electrical energy to satisfy the country's entire needs. Obviously, it would take a little time to develop that, but the natural energy potential is there. That potential is very predictable. Weather forecasting can tell us pretty accurately what type of waves we will get. However, tidal streams are wonderful. They are totally predictable and operate like clockwork. One knows exactly what the flows will be at any given time and at any point around our coasts.'*

He added that there would be minimal environmental impacts with wave or tidal stream plants, and that *'both technologies are ready to begin the process of exploitation. Their costs are about 5p per kW hour, but it is predicted that they could decrease to less than 2p per kW hour when the technologies are fully exploited. The history of technological development suggests that would be no more than an average performance profile for a new technology, so we can probably depend on those figure.'*

He concluded *'Given the order of ascendancy of the availability of natural resources in this country, it is extraordinary that we should be funding wave and tidal power in inverse proportion to its availability. We must urgently revise those priorities and do all that we can to bring on wave and tide.'*

Certainly some device teams have claimed that they can get prices down relatively quickly, if given the right support, with talk of getting down to 3–4 p/kWh for serially produced devices and then lower once the commercial market expanded. And, in similar vein, in its 2006 review, the Carbon Trust used learning curve analysis to identify 'learning rates' for these technologies, based on the fact that costs fall as experience in installing increasing amounts of capacity grows and as economies of volume production are achieved. The slope of the so-called 'learning curve' produced when a kWh price reductions are plotted against kW capacity installed, with both put on logarithmic scales, was 10–15% for wave and 5–10% for tidal current technology.

On this basis, they suggested that, for wave devices, assuming a 15% learning rate from an initial cost of around 22 p/kWh, costs might reach 2.5 p/kWh, competitive with current gas fired combined cycle turbines, by the time around 40 GW of offshore wave capacity was in place. Clearly that is likely to be some way off. More optimistically, they suggest that the same result could occur at around 10 GW of capacity, even given only a 10% learning rate, if the initial starting costs was lower, at 10 p/kWh, due to 'step change' technical breakthroughs. Moreover, they suggest that the 2.5 p/kWh level could be attained at an even lower capacity if a breakthrough was also combined with a 15% learning rate. By comparison, for tidal stream devices, the 2.5 p/kWh point might be reached when only 2.8 GW of capacity was in place (Carbon Trust, 2006).

The key to cost reductions of this order is of course the right sort of support for innovation and development, something that has been lacking in the past and arguably is still only patchy at present. UK Ministers have certainly been faced with challenges from those who felt their rhetoric did not live up to their practice, and who wanted more funding for these technologies. For example, at the hearings of the House of Commons Environment Audit Committee on the report of the Performance and Innovation Units Energy policy report in April 2002, the Committee focused on what they saw as the unfortunate lack of support for wave and tidal power – at that point only around £5 million had been allocated. Asked why he was allocating less money to these projects than to others, the then Energy Minister Brian Wilson, argued that it was because, '*it is demand led*' adding '*I do not think there is any of these technologies where people are queuing up with projects and we have underestimated what we need to invest in order to back these projects. Roughly, the programmes which we have are in proportion to where the demand is coming from, and if that changes, then the programmes can be changed. If there is a technology which is clearly emerging with great potential then we will be right in there and backing it.*'

Subsequently, however, more potential projects must have been found because, in 2004, the DTI allocated £50 million to support work on wave and tidal current technology, with, in 2005, details being released of how £42 million of it would be provided for grants (of up to 25% of the capital cost) and revenue support (around 10 p/kWh for five years). Given that projects can also claim support *via* the Renewable Obligation, at up to 5 p/kWh, it should help more projects to get started (DTI, 2005).

The prospects certainly do now look better, and some see wave and particularly tidal power as likely to overtake wind power. Dr. Abu Bafr Bahaj, who is head of the Sustainable Energy Research Group at Southampton University, has argued that '*The prospects for energy from tidal currents are far better than for wind because the flows of water are predictable and constant. The technology for dealing with the hostile saline environment under the sea has been developed in the North Sea oil industry and much is already known about turbine blade design, because of wind power and ship propellers. There are a few technical difficulties, but I believe in the next five to 10 years we will be installing commercial marine turbine farms.'* (Bahaj, 2003)

However, it is wise to be cautious about the likely scale of the success, unless policies change. The Marine Renewable Energy global markets report produced for ReFocus by consultants Douglas-Westwood Ltd in 2005, estimated that, although globally, installation could soon be running at around 10 MW p.a. on current trends, by 2008 the UK might only have 14.7 MW of wave energy generating capacity in place and around 17.4 MW of tidal current capacity (ReFocus, 2005). That may of course be pessimistic. After all, on present plans, by 2010 it could be that the ten Pelamis Wave Hub project off Cornwall, the 77 MW Wave Dragon project off the Wales and MCTs various 10 MW Seagen tidal farm projects, could all be in place, or well on the way to being commissioned, as could other projects, including the Tidel twin rotor system. In addition there are several projects underway in Scotland, backed by the Scottish Executive as part of its ambitious renewables programme. For example, in 2007 it provided £13 million for nine wave and tidal current projects, including a 3 MW Pelamis wave energy project. So optimistically, the UK might hope to have well over 100 MW of wave and tidal current capacity in place by 2010, or soon after, with expansion well beyond that to be expected subsequently. Even so, as noted earlier, the Carbon Trust looked to only a 3% contribution to UK electricity supply by 2020 and it claimed that 2.8 GW was the maximum total economic potential for tidal current technology (Carbon Trust, 2006).

Given these relatively small projections, it is perhaps not surprising that some have looked back, maybe somewhat nostalgically, at the very large potential of tidal barrages – in theory if all the UK's sites were developed, they could supply a total of around 53 TWh p.a. – about 15% of UK electricity. Certainly enthusiasts for projects like the Severn Barrage have continued to press their merits. In its study of energy options up to 2050, The Royal Commission on Environmental Pollution included scenarios with the Severn Barrage providing 2.2 GW of average rated power (RCEP, 2000). And in 2005 a study published by the UK Institution of Civil Engineers claimed that, far from having adverse environmental impacts, the Severn barrage would actually improve what was it said a relatively barren and inhospitable estuarine environment (Kirby and Shaw, 2005). A new look at barrages was also proposed in the government's 2006 Energy Review.

However costs still remain high and there seems little prospect of significant reductions since, unlike wave and tidal current systems, this is mature, hydro-based, technology. Perhaps more importantly, the emphasis in energy system choice has moved away from very large, inflexible and capital intensive projects like Barrages, to smaller modular devices which can be installed on an easier-to-finance piecemeal basis, and can supply power on a more decentralised basis, meeting local needs more efficiently. For example, despite costing an estimated £10–£15 billion, the 8,600 MW Severn Barrage would only be able to supply full power occasionally – the tidal cycles, although predictable, would not always match the cycles of demand. What use is 8,600 MW in the middle of a summers night?

Of course if and when we have large scale energy storage facilities (including possibly hydrogen storage), then large Barrages could be strong candidates for supplying the power, while tidal lagoons might be an even more economically viable option, with less environmental impacts. It has been claimed that if all the potential UK sites were developed, specially constructed tidal lagoons might supply up to 8% of UK electricity. Moreover, it is also argued that, since, with segmented basins, they can also have some capacity to store water from high tides, so as to better match power output and demand, lagoons should be developed immediately (FoE Cymru, 2004). The lagoon concept is still speculative, and has so far been down-played by the Department of Trade and Industry, who claim that, given the long boundary walls amongst other problems, costs will be high (DTI/WDA, 2006). However, enthusiasts point to a 2004 study carried out for Tidal Electric by W.S. Atkins, who calculated the capital cost at £1.3 million/MW, which they claim confirmed that the proposed Swansea Bay scheme could generate electricity at 3.4 pence/kWh (Atkins, 2004).

## Conclusion

Wave and tidal power systems are developing relatively quickly as new energy options. Some projects, notably the Pelamis wave device and the Seagen tidal current turbine, have reached the commercial application stage, and there are around 60 other wave and tidal current projects of various types under test around the world, about a third in the UK. Wave energy initially took the lead, but tidal current power is now catching up, and is in some ways overtaking, possibly because the technology is similar to that developed for wind turbines. Tidal barrages and lagoons are also based on established technology, but there are operational drawbacks, not least because they basically operate on the head of water captured from high tides twice every 24 hours. Although tidal barrages can in theory be operated in a two-way mode, on the incoming flow as well as the ebb, so as to allow power to be produced over a longer period, two-way turbines are more expensive and suffer more stress. As noted earlier, lagoons can have segmented basin areas for storage of heads of water, and this idea has also been suggested for barrages. However it adds to the cost. More generally, what would in effect be low-head pumped stage systems would not seem to have major advantages over conventional land-based ways of providing pumped storage capacity, with reservoirs with high heads of water.

By contrast, tidal current turbines have the major advantage of being able to operate easily on the ebb and the flow, four times every 24 fours hours, by swivelling around to face the direction of tidal current. The result is that, although as with barrages and lagoons, the power levels would still be low during the lunar neap tide cycles, the daily power outputs can be more nearly continuous. Moreover, if we have many tidal current projects located at different points around the coast, the time variation in the tides cycles at each point would further smooth the combined output. To be fair, that of course would also be true if we had a series of barrages and/or lagoons. And indeed wave devices, which have a completely different wind-determined energy generation profile, unrelated to tidal cycles, could also provide some balancing inputs.

As David Milborrow argues in Chapter 2, we could usefully balance the variable power from other renewables like wind, by using wave and tidal current devices. But we could also perhaps provide some balancing inputs from barrages and/or lagoons.

Certainly lagoons will have lower environmental impacts since they do not block estuaries, and there may be locations and contexts where tidal barrages are suitable – they are after all very site specific. South Korea has been looking at the possibility of locating a 252 MW plant in Sihwa Ho, a large tidal lake in Gyeonggi province, bordering the West Sea, where there is a 5.64 metre tidal range.

However there remain concerns about the potential environmental impacts of large barrages. As noted earlier, in some sites some of these impacts may be positive, but, in effect, blocking major estuaries is likely to have serious ecological implications in many locations. By contrast, although care obviously has to be taken to make detailed assessments, tidal current systems and wave energy systems seem likely to have much less impact – they do not significantly block the flow of tides or waves.

There may be operational issues, such as fouling by discarded nets or other flotsam, and no one denies that the offshore environment is a harsh one. However the UK, amongst others, has extensive marine and offshore engineering experience to call on to develop reliable technologies that can operate in this context, and the incentive to exploit these new offshore energy resources in growing, just at the point when the longer-term prospects for North Sea oil and gas resources look set to decline.

Sea power clearly has a major potential. Whereas until recently we have looked to the sea bed as a source of fossil energy, it may well be that the seas themselves may begin to provide a clean and renewable replacement. The UK Governments Foresight panel has suggested that extracting just 10% of the renewable energy available in the worlds' oceans would provide five times more power than is currently used globally. In terms of marine renewables, there may still be a major gap between theoretical possibilities and practical realities, but wave and tidal current power technologies are developing rapidly, as may tidal lagoons and, possibly, barrages. How quickly they can follow and how much they can deliver will depend not just of the success of technologists in meeting the practical challenges, but also on the level of financial and institutional support they are given. Moreover, as we have seen, the UK is not the only country in the field. The US Aqua Energy Group has suggested that '*offshore wave power has the potential to satisfy 5% to 10% of total US power demand within 20 years*', and similar potentials exist for wave and tidal power around the world. But regardless of whoever takes the lead, it seems inevitable that we will soon be getting power from the sea.

## References

Atkins (2004) 'Feasibility Study for a Tidal Lagoon in Swansea Bay', report for Tidal Electric, Atkins Consultants Ltd., Epsom, September.

Bahaj, A.B. (2003) *Guardian* report on tidal energy 10/02/03.

Carbon Trust (2006) 'Future Marine Energy: Results of the Marine Energy Challenge: Cost competitiveness and growth of wave and tidal stream energy', The Carbon Trust, London.

DTI (2005) 'Marine Renewables: Wave and Tidal-stream Energy Demonstration Scheme', Department of Trade and Industry, London, Jan.

DTI/WDA (2006) 'Tidal Lagoon Power Generation Scheme in Swansea Bay', A consultancy report on behalf of the Department of Trade and Industry and the Welsh Development Agency.

DTI/Carbon Trust (2004) Renewables Innovation Review, Department of Trade and Industry and the Carbon Trust, London.

Elliott, D. with Taylor, D. (2000) 'Renewable Energy for Cities: Dreams and Realities' paper to the UK-ISES Conference on 'Building for Sustainable Development' May 26, RIBA, London.

Elliott, D. (2004) 'Tidal Power' in G. Boyle (ed.) '*Renewable Energy*' OU/Oxford University Press, pp. 196–242.

FoE Cymru (2004) 'A Severn barrage or tidal lagoons? A comparison Briefing', Friends of the Earth Cymru, Cardiff, January http://www.foe.co.uk/resource/briefings/severn_barrage_lagoons.pdf.

Kirby, R. and Shaw, T. (2005) 'Severn Barrage UK – a environmental reappraisal', *Engineering Sustainability*, 158, EC1, Proceedings of the Institution of Civil Engineers, March, pp. 31–9.

ETSU (1999) 'New and Renewable Energy: Prospects in the UK for the 21st century'. ETSU report R122, Energy Technology Support Unit, Harwell.

MPS (2006) Modern Power Systems, May 25th, June 28th and Nov 25th and Nov 29th web news reports: www.modernpowersystems.com.

RCEP (2000) 'Energy – the changing climate', Royal Commission on Environmental Pollution, Twenty-second report, Cmnd 4749, HMSO, London.

ReFocus (2005) Refocus Marine Renewable Energy Report: Global Markets, Forecasts and Analysis 2005–2009, Douglas-Westwood Ltd, Elsevier, London.

Ross, D. (2002) 'Scuppering the waves: how they tried to repel clean energy', *Science and Public Policy*, Vol. 29, No. 1 Feb., pp. 25–35, DTI quoted on p. 34.

Select Committee (1984) House of Commons Select Committee on Energy, Session 1983–84, Ninth Report, 'Energy R, D&D in the UK', HMSO, London, p. xxxii.

Select Committee (1992) House of Commons Select Committee on Energy, Session 1991–92, Fourth Report, 'Renewable Energy', Vol. III, March, HMSO, London, p. 125.

Select Committee (2001) House of Commons Select Committee on Science and Technology, Session 2000–2001, Seventh Report, 'Wave and Tidal Energy', HC 291, April, HMSO, London, p. iv.

Smith, D. (2005) 'Why wave, tide and ocean current promise more than wind', *Modern Power Systems*, Part II-European development (Aug) Part III – The Americas (Dec).

Swanturbines (2006) OXERA market forecast scenario relayed on Swanturbines web site: http://www.swanturbines.co.uk/.

Welsh Affairs Committee (2006) Uncorrected transcript of evidence to the House of Commons Welsh Affairs Select Committee on March 28th 2006. Final version to be published as HC 876-iv.

# 5

# Solar Power: Using Energy from the Sun in Buildings

*Susan Roaf and Rajat Gupta*

Solar Energy is the easiest form of renewable energy to integrate into the fabric of a building, and in turn a city, and is capable of providing a significant amount of the necessary electricity, heat and hot water for the comfortable operation of a building over a year. Over the last decade there has been a huge growth in large-scale solar generation plants using a range of technologies including Solar Thermal Power Plants, Parabolic Dishes, Solar Dishes, Solar Trough Farms, Solar parks and Solar Power Towers (Schlaich, 1995; Markvart, 2000) in large arrays. Solar energy must inevitably become the most widely used form of renewable energy for buildings and cities over the next decades, even if it is not the most energy productive of the renewables. This is for many very good reasons:

1. **Building integration: Renewable energy as a design feature**
   Solar energy is the easiest form of renewable energy technology (RE) to integrated physically into a building. It can be a driver for building form, as with passive solar sun spaces and collectors, and as building material, forming a rain screen cladding to the external envelope of a building (Gadsden, 2001, p. 23; Yannas, 1994, pp. 12–13). Photovoltaic (PV) systems and passive solar elements in particular can be an attractive architectural feature for designers. In addition solar technologies do not have the same problems of reverberation when coupled with a building structure as do wind generators and neither do they produce smoke as a system by-product as do fossil fuel burners, making their wider use in cities unacceptable. Solar heat collectors may need a small amount of power for the pump, but they do not require large amounts of electricity to run as do ground coupled heat pumps, and PV solar needs no power input.

Solar technologies are clean, quiet, and robust to operate and can result in avoided costs of construction if used as rain screen cladding. They may require occasional cleaning over time depending on the climate in which they are located.

2. **High level of public acceptability for solar technologies**
Studies by Chadwick *et al.* (2002) to gauge the acceptability of different renewable energy technologies by the local community, as perceived by planners, reveal that in comparison to large-scale wind installations and waste incineration, solar energy technologies in the form of passive solar, solar PV and solar heating are likely to be favourably viewed by local people (Chadwick *et al.*, 2002, pp. 3–4). See Figure 5.1. The use of solar design in buildings since the earliest times is linked with an embedded cultural appreciation of the power and importance of the sun to human civilisations over millennia (Knowles, 2006).
3. **Point of use energy generation**
The ability to integrate solar technologies into the form and envelope of a building also has the added advantage of ensuring that the energy is generated at point of use, thus minimising supply line energy losses during the process of delivering the energy to the building. The total amount of energy wasted by delivering it to the building is enormous. The electric utility sector in the US produces over 2.0 billion metric tons of carbon dioxide emissions per year – slightly more than one-third of all US carbon dioxide emissions. Among US utilities, energy losses from transmission and distribution typically are in the range of 5% to 10% and average about

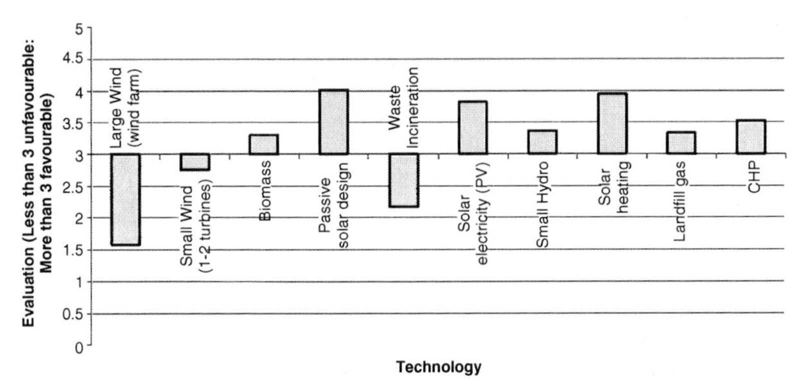


*Figure 5.1* Acceptability of renewable energy technologies by the local community, as perceived by planners

*Source*: Chadwick *et al.*, 2002, p. 4.

7% (http://www.eia.doe.gov/oiaf/1605/vr98rpt/chapter2.html). The carbon dioxide emissions associated with the lost energy total around 150 million metric tons per year (based on the average fuel mix for the United States).

4. **Non-polluting technologies**
   Solar technologies have low pollution impacts. There will of course be emissions from producing the energy used to manufacture the solar systems, and this is relatively high for solar PV, but even with PV the energy debt will be paid back within between one to three and a half years (http://www.nrel.gov/ncpv/energy_payback.html), and the pollution resulting from the sealed units is non-existent, although concerns about some of the substances used in the production of the panels have tended to drive the markets away from certain panel types including those made with the carcinogenic cadmium telluride (Fthenakis, 2003).

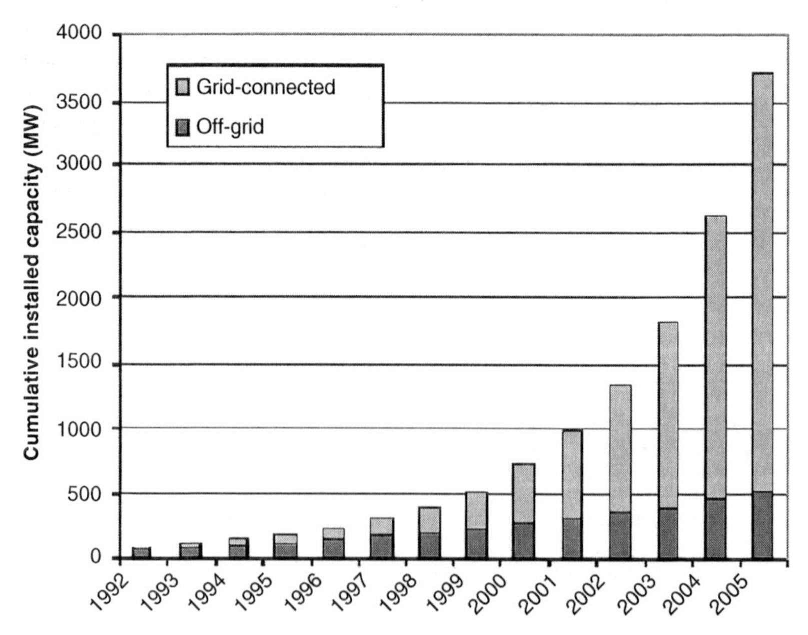


*Figure 5.2* Cumulative installed PV power MW globally, as of the end of 2005. The market grew by 680 MW in 2004–2005

*Source*: IEA-PVPS website, www.iea-pvps.org. A full review of photovoltaic markets can be seen on the International Energy Agency website on photovoltaics: http://www.iea-pvps.org/index.html

5. **An easy to use technology**
   Solar systems, at the domestic level in particular, are accessible to the ordinary building owner and designer, being neither difficult to understand or implement. They have been relatively widely used since the energy crisis of the 1970s and since then numerous books, courses and demonstration projects have publicised the benefits and mechanics of building integrated solar energy (Brown and Dekay, 2000; Mazria, 1979; Muneer, 2004; Schafer, 1996; Szokolay, 2003). Once fiscal incentives had been developed in Germany with their ground breaking 1,000 solar roof project in the 1990s, the uptake of the technology has been rapid and Germany leads Europe in terms of installed systems per capita and in terms of developing a healthy Low Carbon Economy based on the production and implementation of solar technologies (Capra and Pauli, 1995). See Figure 5.2.
6. **Load shaving: Peak generation comes at times of peak demand**
   One of the greatest challenges we have in fighting climate change is to reduce our dependency on air-conditioning. It is rapidly becoming the worlds number one source of greenhouse gas emissions as countries where the peak load on the grid typically occurred during the winter heating season are now shifting to strong summer peaks in their electricity demand, as a result not only of the warming climate but also of increasing poor standards of climatic design in over-glazed modern buildings (Roaf *et al.*, 2005, pp. 213–39).

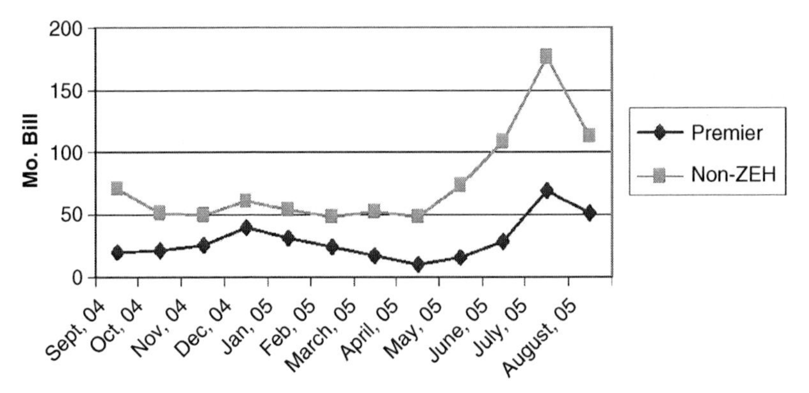


*Figure 5.3* The difference between Summer monthly peak electricity use, in terms of monthly bills, in ordinary homes in Sacramento, California and 100 'Premier' homes showing the significant impact on summer peak use of having solar hot water and photovoltaics integrated into energy efficient homes, called ZEH – 'Zero Energy Homes'

*Source*: Bruce Baccei.

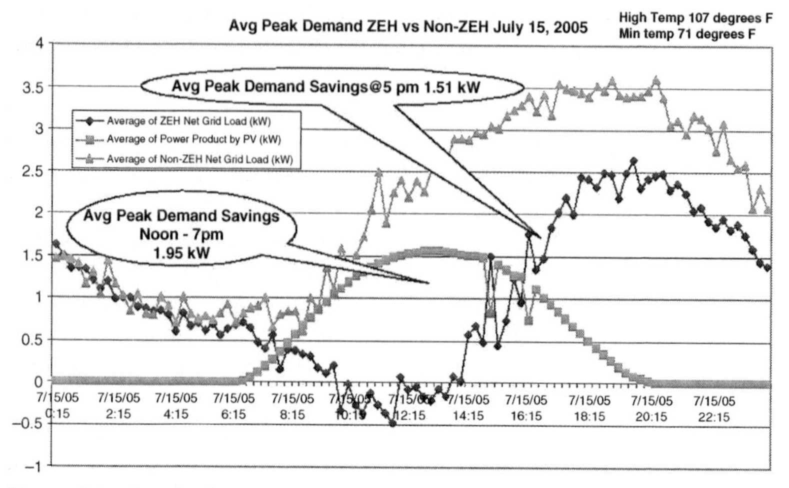


*Figure 5.4* Graph showing the significant daily peak load shaving that is affected by a 2 kWp PV system on the home, in kW

*Source*: Bruce Baccei.

Figures 5.3 and 5.4 show the significant monthly and daily load shaving impacts that can have been achieved in the Premier Zero Energy Homes development in Sacramento California where each of the 100 houses has a 2 kW AC solar electric home power system, a tankless water heater that produces 6.5 gallons of hot water per minute, a mechanically designed heating and air conditioning system, spectrally selective glass windows, and tightly sealed air ducts (http:// www.builtbypremier.com/P_meadows/news_zero-energy.html).

In many areas of the world where the warming climate is increasingly coupled with rising air-conditioning costs and rapidly rising electricity prices, the need for affordable and effective solar cooling technologies is urgent, and many recent developments in the field are already being implemented in buildings. It is not only individuals and companies who are challenged, but governments and regions are also pushed towards more energy supply blackouts. In the increasingly privatised energy supply sector, a generic problem is the lack of investment in 'spare' generation capacity, so in extreme weather events, arising from either the heat or cold, electricity generation and supply systems around the world are failing at times of peak demand (Roaf *et al.*, 2005, pp. 269–303). Particular groups who are disadvantaged in this vicious circle include the millions of Americans who live in thin skinned, poorly insulated tract houses with very little thermal mass, kept comfortable in often extreme climates, with energy hungry, heating and cooling systems. The point will come soon enough where ordinary families simply cannot afford the electricity to keep them cool or warm enough to be safe and the common sense of simply using solar heating or solar cooling systems, free to run after the initial capital investment, will become increasingly apparent in a warming world with increasingly scarce fossil fuels.

7. **Load shoring**
   The need to plan energy generation and transmission capacities in the face of growing demand and aging plants and delivery systems is a major concern. In areas where the transmission lines are old but replacement costs are prohibitive, or where distributed demand is placing excessive loads on local supply networks, the use of end of line, embedded renewable generation capacity can shore up a weak line, reinforcing the supply quality and reducing the need to replace aging supply lines. Point of use renewable energy generation capacity can also be planned to reduce the need to replace the centralised conventional generation plant. This reasoning is being widely used in areas such as California, where the grid supplies remote locations and the age of the system is an issue. In addition, the growth of remote demand and new supply requirements favour at least an element of renewable generation in the mix. Where this is in remote settlements, solar energy systems are often the preferred generator.
   http://www.energy.ca.gov/2004_policy_update/index.html#endnotes
8. **Low maintenance technology**
   Once solar systems are fixed and commissioned on a building they are typically very robust and quite simply work unless a failure occurs in one of the related mechanical or electrical systems. They require little cleaning (about the same as for windows) and do not suffer from the necessity for regular maintenance and replacement of parts as wind systems are liable to do. The simply sit on a building and work.
9. **A technology for social transformation and resilience**
   The decades ahead will be socially extremely challenging as gas and electricity bills rise and more people slip into fuel poverty, exacerbated by extreme weather events that require more energy to combat their impacts. It is the poor and the physically vulnerable who are most at risk and early investment in building integrated solar systems is perhaps the most effective method of 'future-proofing'

citizens and their communities against these changes. Once the initial capital investment in solar technologies has been made they simply sit there and work to significantly reduce the energy costs of a building, particularly when used in conjunction with energy efficiency measures. Once installed, the cost of the energy cannot rise, unlike that from a conventional or nuclear power plant.

The case study below of the Oxford Ecohouse demonstrates this and reinforces the links between increased energy awareness in the home and the use of solar systems. The follow-on Oxford Solar Initiative demonstrates how local solar action campaigns can bring the community together to increase individual resilience, and also community preparedness for climate changes and high energy prices.

Unlike conventional sources, solar technologies offer the opportunity for:

- Energy awareness training of households
- Reduced energy running costs of homes and carbon emissions from them
- Community level environmental campaigns building group resilience and preparedness for climate and energy price changes
- The taking of responsibility for emissions at an individual and local level

By contrast the concentration of generation capacity is associated with the:

- Potential for catastrophic failure of energy supply systems on a regional scale
- Rising energy bills over which individuals have no control
- Disempowerment of people as potential agents of change

Solar Energy can therefore be seen as an effective driver for the implementation of the perceptional changes necessary in society at large to engage people in becoming part of the solutions to the enormous challenges ahead.

## Functions of building integrated solar systems

In the UK, each square metre of a south-facing roof receives around 1,000 kWh of solar radiation during a year. This means that the roofs of many homes receive more energy from the sun in a year than is needed to provide both the space and water heating. Over the last two decades a generation of pioneering solar homes and architects have sprung up around the world (Strong and Scheller, 1993; Galloway, 2004; Kachadorian, 1997; Roaf *et al.*, 2003), to disprove the myth that 'there is not enough sun in northerly latitudes' to make solar energy worth the investment. The idea that solar technologies are more cost effective in the lower latitudes with their higher levels of solar radiation has been dispelled in the work of Colin Porteous who makes clear that at higher latitudes the importance of solar energy can be greater to building users as there is more need for the heat they produce than in already hotter lower latitudes (Porteous, 2005).

There are a number of ways in which solar energy can be used in buildings. Enough solar radiation falls on the roofs of most UK homes to supply a significant proportion of the domestic energy needs of that home. Solar energy can be used in homes for:

1. *Daylighting:* Careful design of a building to introduce sunshine or daylight into as many areas of a home as possible will significantly reduce the need for energy use for artificial lighting. The use of internal sources of natural light from internal windows, fanlights, skylights and plenum lighting can save a significant proportion of the annual lighting load of the house (Crisp *et al.*, 1988).
2. *Passive solar heating*: including direct solar gain that reduces space heating demand by heating the air in the rooms exposed to the sun, and the walls of those spaces. The resulting radiant and convective solar heating can be used to warm non exposed areas of the building and to facilitate clothes drying in well ventilated sunny spaces (Solar Energy Applications Laboratory, 2005; Kachadorian, 1997; Galloway, 2004).
3. *Active solar heating*: typically involves the use of solar hot water panels linked by a thermo-syphonic or pumped water system to the hot water tanks of the building (Roaf *et al.*, 2003).
4. *Solar electric systems*: Solar electric (photovoltaic) panels in the UK generate electricity from sunlight and produce most power when tilted at around 30–40° to the horizontal and, in the northern hemisphere, positioned roughly facing south (south west to south east). The energy used in the manufacture of PV can be recouped within three to four years – a small proportion of the 30 years for which it will generate electricity (IEA PVPS, 2002; Scoones, 2001, p. 13). Nowadays, PV cells can be made of a range of materials, which include monocrystalline silicon, polycrystalline silicon, amorphous

silicon, cadmium telluride (CdTe), copper indium diselenide (CIS), gallium arsenide and gratzel cell (Germans Solar Energy Society, 2005). Of these, the PV cells made from monocrystalline silicon are said to be the most efficient, converting 15% of the solar radiation into electrical energy (Edwards and Turrent, 2000, p. 52; Jardine and Lane, 2003, pp. 6–8). Typical efficiencies of PV technologies are given in Table 5.1 below:

***Table 5.1*** **Efficiencies of different PV technologies**

| Technology | Efficiency (%) |
|---|---|
| Monocrystalline silicon | 12–15 |
| Polycrystalline silicon | 8–12 |
| Amorphous silicon | 4–6 |
| Cadmium telluride | 7 |
| Copper indium diselenide | 9 |

*Source*: Jardine and Lane, 2003, p. 9.

5. *Dehumidifying solar air-conditioning systems*. There are an emerging range of technologies with roof top solar collectors used to power desiccant and absorption components that directly remove the water vapour (latent heat) from the air passing through the cooling system and can be used to control the temperature and humidity of the air in the system (Henning, 2003; http://www.iea-shc.org/task25/index.html)
6. *Solar chimney convective cooling systems*. For millennia people in the Middle East have been convectively cooled by air streams driven by solar chimneys at the hottest times of year when pressure driven ventilation systems fail in high pressure weather systems. In recent decades significant work has gone into developing solar chimneys that use the mass of the building to cool the air inside the house and using a stack flow draw it through living spaces, up and out creating an air stream that convectively cools the building occupants and in many cases, uses the cool night air circulated in such systems to re-cool the fabric of the building itself (http://en.wikipedia.org/wiki/Solar_chimney; Beazley and Harverson, 2002).
7. *Rainscreen cladding*. Building integrated PV systems are now often extremely sophisticated in their construction and balance of system design and are increasingly used as an innovative design features such as rain screen cladding for roofs and building walls in urban, suburban and rural locations (Solar Energy International, 2004).

## The Oxford Ecohouse: A case study of a UK solar house

The Oxford Ecohouse was completed in April 1995 with the first PV roof in the UK. This high profile house was extensively monitored and written on, and demonstrated to the general public the combination of energy efficiency, solar energy, biomass and first class passive solar design homes could not only produce a building with very low energy bills and a high standard of comfort internally, even in extreme weather, but also showed that with 1990s technology a reduction of up to 90% $CO_2$ emissions could be achieved with no reduction of Quality of Life for the occupants. See Table 5.2.

The house is laid out with rooms arranged around a central core incorporating a service duct, stairs to the first floor and hallways that lead on the ground floor to the front entry porch. Bathrooms are positioned over the kitchen to reduce pipe runs, and hence material use and travel times for hot water. The front and back doors are protected by buffer spaces, a porch to the north and a two story double glazed conservatory to the south, with a balcony on the first floor between two bedrooms.

## Key characteristics of the Oxford Ecohouse

- Solar radiation: 1,000 Kwh/m$^2$ per year
- Heated volume: 468 m$^3$
- Heated floor area: 233 m$^2$
- Sun space floor area: 12.6 m$^2$
- Thermal characteristic: the heat loss from occupied zones is 0.84 W/m$^2$C
- Measured air change rate per hour before modification is 0.42
- Wood burning high mass stove or kacheloven
- 5 m$^2$ flat plate solar collector calculated using KEW weather data, predicted to supply 77% of the household requirements (Viljoen, 1995)
- 4 kWp mono-crystalline BP Solar 585 Saturn PV array, still generating over 3,000 kWh/pa

The house is situated in an area that was calculated to receive only approximately 4 peak sun hours in summer, and 0.6 peak sun hours in winter. The variation in output from winter to summer is therefore significant and the array size has tended to be specified in order to achieve a reasonable level of autonomy for the house in winter. Power

***Table 5.2*** **Comparison of annual gas and electricity use in three semi-detached homes built to typical 1995 Building Regulations Standards, Zero $CO_2$ and Zero Energy Standards with the $CO_2$ emissions from the Oxford Solar House. The Oxford Ecohouse is 232 $m^2$ and the three bedroom semi-detached house is 162 $m^2$ in size**

| | **Typical** | **Zero $CO_2$** | **Zero** | **Oxford Ecohouse** |
|---|---|---|---|---|
| Space heating (kWh) | 7,926 | 3,172 | 240 | 4,500 |
| Hot Water (kWh) | 4,548 | 2,319 | 1,660 | 500 |
| Cooking (kWh) | 656 | 330 | 330 | 500 |
| **Total annual gas use (kWh) – (A)** | **13,130** | **5,821** | **2,230** | **5,500** |
| **$CO_2$ emissions ($kgCO_2$) for annual gas use (× 0.19 $kgCO_2$/kWh) – (B)** | **2,495** | **1,106** | **424** | **1,045** |
| Lights and appliances (kWh) | 3,000 | 2,700 | 2,100 | 3,500 |
| Pumps and Fans (kWh) | 175 | | 200 | |
| **Total annual electricity use (kWh) – (C)** | **3,175** | **2,700** | **2,300** | **3,500** |
| Electricity imported to Oxford ecohouse | | | | 2,500 |
| $CO_2$ emissions ($kgCO_2$) from imported electricity (×0.43 $kgCO_2$/kWh) | 1,365 | 1,161 | 989 | 1,075 |
| PV electricity made at Oxford Ecohouse (kWh) | | | | 3,000 |
| PV electricity exported from Oxford Ecohouse (kWh) | | | | 2,000 |
| Avoided $CO_2$ emissions ($kgCO_2$) for exported PV electricity* | | | | – (860) (1075–860) |
| **Net $CO_2$ emissions ($kgCO_2$) from annual electricity use – (D)** | **1,365** | **1,161** | **989** | **215** |
| Total annual energy use (A +C) kWh | 16,305 | 8,521 | 4,530 | 9,000 |
| Total annual $CO_2$ emissions (B+D) ($kgCO_2$) | 3,860 | 2,267 | 1,413 | 1,260 |
| Area of the house ($m^2$) | 162 | 162 | 162 | 232 |
| Annual $CO_2$ emissions per $m^2$ ($kgCO_2/m^2$/yr) | 23.8 | 13.9 | 8.7 | 5.4 |

*NB*: It is the solar generated electricity that makes the greatest contribution to reducing $CO_2$ emissions from the Oxford House.

* exported PV electricity is used in adjacent buildings replacing imported energy form the grid so $CO_2$ for this displaced electricity is deducted from the house total $CO_2$

has to be drawn from the utility during winter days and also during night-time. During the summer months, energy surpluses were predicted to be around 12 Kwh per day which is greater than the house energy deficit in winter. The house therefore is expected to have a positive energy balance. In reality, eleven years after the system was commissioned the system produces around 20 kWh a day during the summer months and up to 12 kWh in the spring and autumn months and is still producing over 3,000 kWh a year. Over the last 3 years electricity usage has averaged around 4,000 kWh electricity a year considerably up from the original estimate of 3,000 kWh a year, due largely to the increase in electrical goods in use.

The PV system generates 80–85% of the electricity consumed in the house, but only around 25% of the PV generated electricity is used in the house. The remainder is exported to the grid. Differential rates apply to the purchase and sale of the electricity necessitating the paying c. 10 p a unit for imported electricity and being paid 4 p a unit for every unit sold. The gas used in the house is currently around 6,000 kWh per annum.

The Oxford house has been visited by thousands of students, home owners, designers and politicians who have learnt from it that solar energy is 'not rocket' science, that it is robust and works well to significantly reduce household energy bills and carbon emissions from a very comfortable home. The confidence that this single project instilled in the people of Oxford provided a solid foundation for the Oxford Solar Initiative.

## The Oxford Solar Initiative

The Oxford Solar Initiative emerged from a new generation of research and development that seeks city-wide applications of renewable energies and other means of greenhouse gas emissions reductions and absorption that will be applied in a coherent spatial and social context, as well as within community-wide framework (see: http://oxfordsolar.energyprojects.net).

The Oxford Solar Initiative proposes three areas that are being of focus. They are to be advanced simultaneously. These are:

(a) **$CO_2$ reduction focused urban planning strategies**
In part due to restrictions imposed by clients, construction, design, architecture and development firms can, but rarely do, have a major influence over the embedded carbon in new developments.

Oxford City Council now gives permission automatically for solar arrays except in conservation areas where planning permission must be sought. They also have renewable energy targets embedded in their Planning Guidance documents (see: http://www.oxford.gov.uk/news/news.cfm/environment/2/item/552)

(b) **Targets, baseline studies and scenario development**
*Oxford Council has set a emissions reduction target of 3% $C0_2$ as part of its 2005 Climate Change Action Plan and its Natural Resource Impact strategy* www.oxford.gov.uk/files/meetingdocs/30243/item7.pdf *and* http://www.oxford.gov.uk/planning/nria-spd.cfm

(c) **Urban energy technologies, industry and business development**
Having seen the potential for development of a Low Carbon technologies a number of groups in the city are working to push forward this agenda.

The overall objective of the Oxford Solar Initiative was to find the best ways to introduce Solar Energy Technologies (SET) and the rational use of energy (RUE) in Oxford. The initiative contains several clear objectives. Some of these are:

Goal One: 10% of all houses in Oxford will have solar systems by the year 2010.

Goal Two: To implement a capacity building programme in the local government for the provision of information, training, and other services oriented to $CO_2$ mitigation strategies.

Goal Three: To establish strategic alliances, and participation, of the local government, households, business organisations, energy supply companies and community organisations to fulfil Oxford's $CO_2$ reduction targets.

Goal Four: To initiate and implement a solar campaign to support local $CO_2$ reduction initiatives at every level within the Oxford community from primary school children to business leaders.

The original feasibility study which surveyed 700 householder, was carried out in Oxford to investigate customer attitudes towards installation of PV, solar hot water and/or passive solar features in combination with energy efficiency measures on houses in North Oxford. The results of the survey are shown in Table 5.3. The results showed 91% of the respondents agree or agree strongly that they would consider applying energy efficiency measures. 65% of them also agree to consider using

*Table 5.3* **Results from the questionnaire survey**

| External wall construction | |
|---|---|
| Cavity wall / Not insulated | 4.0% |
| Cavity wall / Insulated | 3.0% |
| Solid brick/stone (mainly pre-war) | 67.4% |
| Don't know | 8.9% |
| Other | 16.8% |

| Draught proofing | |
|---|---|
| All | 12.9% |
| Most | 11.9% |
| None | 25.7% |
| Some | 49.5% |

| Low energy lighting | |
|---|---|
| All | 1.0% |
| Half | 2.0% |
| Most | 9.9% |
| Some | 66.3% |
| None | 20.8% |

| Type of heating | |
|---|---|
| Condensing boiler | 9.9% |
| Standard boiler | 65.4% |
| Standard combi | 14.9% |
| Other | 9.9% |

| Consider applying energy efficiency | |
|---|---|
| Agree | 26.7% |
| Agree strongly | 64.4% |
| Don't know | 8.9% |

| From yes, price to pay for solar hot water | |
|---|---|
| £ 1,000 | 37.0% |
| £ 1,500 | 26.2% |
| £ 2,000 | 9.2% |
| £ 3,000 | 3.1% |
| Don't know | 24.6% |

| Loft insulation | |
|---|---|
| 25 mm (1") | 2.0% |
| 50 mm (2") | 11.9% |
| 75 mm (3") | 10.9% |
| 100 mm (4") | 15.8% |
| 150 mm (6") | 5.9% |
| 200 mm (8") | 2.0% |
| None | 4.0% |
| Don't know | 47.5% |

| Hot water cylinder insulation | |
|---|---|
| Jacket | 34.7% |
| Rigid foam | 36.6% |
| None | 28.7% |

| Secondary/double glazing | |
|---|---|
| All | 25.7% |
| Half | 3.0% |
| Most | 11.9% |
| Some | 40.6% |
| None | 18.8% |

| If standard boiler or combi, year installed | |
|---|---|
| Before 1980 | 37.0% |
| 1980–1990 | 29.6% |
| 1990–2000 | 21.0% |
| After 2000 | 12.4% |

| Consider using solar energy | |
|---|---|
| No | 2.0% |
| Yes | 64.4% |
| Don't know | 33.7% |

| From yes, price to pay for solar PV | |
|---|---|
| £ 1,500 | 44.6% |
| £ 2,500 | 18.5% |
| £ 5,000 | 6.2% |
| Don't know | 30.8% |

*Source*: Roaf et al., 2003.

solar energy. And among these, 41% agree to pay £1,000–£1,500 for solar hot water systems and £1,500–£2,500 for solar PV systems (Roaf *et al.*, 2003).

Today some 450 individual homes in the area have put solar systems on their roofs and the demand is growing. In a Solar Fair held on the 5th April 2006 some 1,500 citizens visited the Town Hall to attend lectures and find out more about the technologies. There are two developing projects to install a community micro-grid on a Solar Street in Harpes Road, Oxford and to make one whole Ward of the city, Wolvercote, a Low Carbon, Solar Village.

## Reducing carbon emissions from Oxford housing using solar energy systems

The issue of system costs to the consumer in relation to carbon emission reductions and the relative cost benefits to the individual, and to the UK government, are dealt with using the DECORUM model. This provides the cost per tonne of carbon saved for the full range of UK house types and can be used to identify a comprehensive range of emission reduction strategies including energy efficiency and solar technologies for a case study area in Oxford. DECoRuM, a GIS-based carbon-counting and carbon-reduction model has been deployed in a case study in Oxford to estimate the baseline energy consumption in, and carbon emission from, existing dwellings on an urban scale (Gupta, 2005a; Gupta, 2005b). The case study area in Oxford comprised 318 dwellings which were representative of the age groups and built forms prevalent in the UK housing stock.

Recent studies by the Department for Communities and Local Government (DCLG) have confirmed that retrofitting a property to the highest energy performance standards usually requires some form of low carbon source, i.e. micro-generation such as heat pumps or solar energy systems (DCLG, 2006). This is why four alternative packages of $CO_2$ reduction measures (energy efficiency measures, low carbon technologies and solar energy systems) were developed to explore possibilities of achieving reductions of over 60% in the Oxford case study dwellings. These include:

*Package 1:* Energy efficiency measures
*Package 2:* Energy efficiency measures + low carbon technologies (GSHP and domestic micro-CHP)
*Package 3:* Energy efficiency measures + active solar energy systems (SHW & PV)

*Package 4:* Energy efficiency measures + low carbon technologies + active solar energy systems

As Figure 5.5 shows, any of the packages, if deployed in the case study dwellings in Oxford, could achieve annual reductions of around 60% or more above baseline emissions. However the cost for deploying any of the packages to reduce a tonne of $CO_2$ emissions lies in a remarkable range of £6–£77 per tonne of $CO_2$ saved, depending upon the package of measures used, and which scenario of capital costs (low or high) is employed. While Package 1, consisting of energy efficiency measures seems to be the cheapest way of reducing a tonne of $CO_2$ emissions in both low and high capital cost scenarios (in line with UK Government policy), package 2, comprising low carbon technologies, saves almost 63% above baseline emissions. However, the cost per tonne of $CO_2$ saved is much lower than for package 3, which consists of energy efficiency measures and active solar energy systems (SHW and solar PV). Nevertheless, low carbon technologies, although efficient, rely on fossil fuels, be it gas for micro-CHP units or electricity to run GSHP systems. That is why, in package 3, energy efficiency measures are adopted along with SHW and solar PV systems.

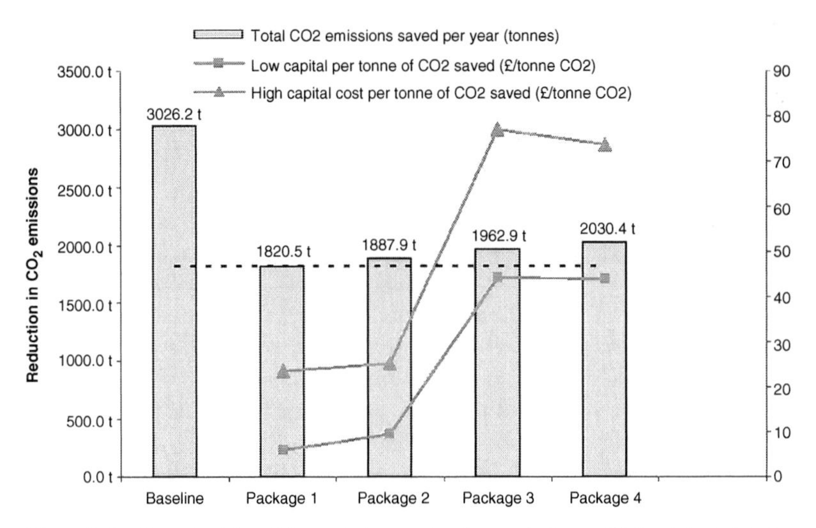


*Figure 5.5* Potential for $CO_2$ emission reductions of above 60% attributable to the four alternative packages deployed in Oxford case study area

*Source*: Gupta, 2005a.

It is important to note that individually, if SHW and solar PV systems are installed in the selected case study dwellings, they cost respectively £335 and £644 per tonne of $CO_2$ emissions saved in a low capital cost scenario. However, when applied in combination with energy efficiency measures in package 3, the cost for reducing a tonne of $CO_2$ emissions drops to £44. Importantly, this indicates that in order to bring down the costs of solar systems and increase their take-up, these should be installed in tandem with energy efficiency measures. Firstly, the total domestic energy should be reduced as much as possible by using appropriate energy efficiency measures, and then the reduced energy demand should be met by solar energy systems.

When SHW and solar PV systems are combined with both energy efficiency measures and low carbon technologies in package 4, they help to achieve potential reductions of around 67% above baseline emissions. Despite the increase in reductions by moving from package 3 to package 4, the cost per tonne of $CO_2$ emissions saved by package 4 is close to the cost in package 3. In summary, these modelled results demonstrate the cost case for investing in energy efficiency before solar technologies but also makes the clear point that in order to achieve the levels of carbon reductions required to meet realistic climate protection targets then there is little choice but to invest in building integrated solar technologies on a wide scale.

## Cost-effectiveness of solar energy systems and increasing fuel prices

The cost case for early investment in solar technologies such as solar HW and solar PV, is further reinforced by modelling the cost benefits of these systems with a doubled and tripled unit cost for electricity and gas in the UK. Based on application of the DECoRuM model in a case study comprising 318 dwellings in Oxford which are representative of the age groups and built forms prevalent in the UK housing stock, the cost-effectiveness of solar hot water (4 $m^2$) and solar PV (2 kWp) systems are evaluated for increasing fuel prices, using a Net annual cost method (NAC). See Figure 5.6. The NAC method is based on costs minus savings, with the result that a cost-effective measure has a negative value for NAC (Iles and Shorrock, 2004, p. 3). A discount rate of 3.5% is assumed for all calculations in this study, in line with treasury guidance (H.M. Treasury, 2003). A measure is taken to be cost-effective if the NAC is negative; the larger its absolute value, the more cost-effective it is. The NAC is divided by the annual $CO_2$ saving to give the *net annual cost per tonne of $CO_2$ saved*, a particularly important measure of cost-effectiveness. Cost-effectiveness is assessed for both solar HW and solar PV systems using both low and high estimates of capital costs extracted from the latest report by BRE (Shorrock *et al.*, 2005). These costs are listed in Table 5.4, indicating what they correspond to.

The fuel prices scenarios assumed are as follows:

1. Current fuel price scenario: 10 p/kWh for electricity and 3 p/kWh for gas
2. Double fuel price scenario 20 p/kWh for electricity and 6 p/kWh for gas
3. Triple fuel price scenario 30 p/kWh for electricity and 9 p/kWh for gas

***Table 5.4*** **Typical low and high capital costs for installing solar hot water and solar PV systems in homes**

| Measure | Capital cost (£) | | Description |
|---|---|---|---|
| | Low | High | |
| Solar hot water system (4m$^2$) | 1,650 | 2,475 | Grant-aided and typical cost |
| Solar PV system (2kWp) | 6,900 | 13,300 | Grant-aided and typical cost |

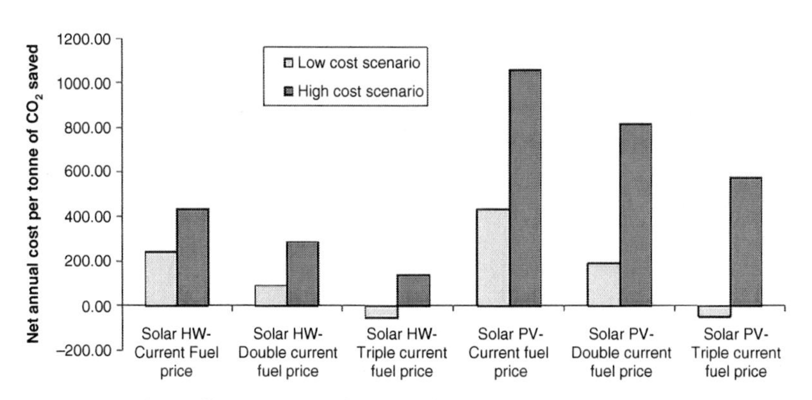


*Figure 5.6* Cost-effectiveness (£/tonne of $CO_2$ saved) of solar hot water and solar PV systems under low and high capital costs for current, double and triple fuel prices, when applied to a case study area in Oxford city. Using Net Annual Cost method

*Table 5.5* Analysis of 2 kWp solar PV systems for increasing fuel prices, using the simple payback period approach

| | Fuel saved | Fuel price | Revenue from ROC | ROC revenue per year (£) | Saving/year from PV electricity (£) | Total saving (£) | Capital cost (£) | | Simple payback (years) | |
|---|---|---|---|---|---|---|---|---|---|---|
| | | | | | | | Low | High | Low | High |
| Solar PV-Current fuel price | Electricity | 10 p/kWh | 4 p/kWh | 60 | 150 | 210 | 6900 | 13300 | 32.9 | 63.3 |
| Solar PV-Double current fuel price | Electricity | 20 p/kWh | 4 p/kWh | 60 | 300 | 360 | 6900 | 13300 | 19.2 | 36.9 |
| Solar PV-Triple current fuel price | Electricity | 30 p/kWh | 4 p/kWh | 60 | 450 | 510 | 6900 | 13300 | 13.5 | 26.1 |

Interestingly, it is seen that for the Oxford case study area, Solar HW (4 $m^2$) and solar PV (2 kWp) become cost-effective (for low capital costs) in the triple fuel price cost scenario. The triple cost scenario may not be very far in the future as average domestic gas prices have changed from 1.4 p/kWh in 2004 to 3 p/kWh in 2006 (>100% increase) whilst average electricity prices have changed from 7 p/kWh in 2004 to 10 p/kWh in 2006 (50% increase).

In addition to this thorough cost-benefit analysis, another way of analysing cost-effectiveness is by calculating the *simple payback* period for each of the measures. See Table 5.5. The UK government recommends that a measure be deemed to be cost-effective if the simple payback period is no greater than 7 years (ODPM, 2004, Section 3, p. 22). Although 2 kWp solar PV systems presently have a high payback period equal to almost 46 years even after including grants. However, from 1 April 2004, UK legislation has enabled micro-generators such as domestic solar PV systems to qualify for Renewable Obligation Certificates (ROCs), and sell these to suppliers (Green Electricity Marketplace, 2004). Assuming an average ROC price of 4 p/kWh as proposed by Watson (Watson, 2002; Watson, 2004, p. 1985) using scenarios developed within the Tyndall Centre, the ROC revenue for 1,500 kWh of annual electricity output by a 2 kWp system is £60/year. This ROC revenue adds to the savings of avoided cost of purchasing electricity and brings down the payback period of PV systems from 46 years in a *low capital cost current fuel price scenario* to 13.5 years in a *low capital cost triple fuel price scenario*. This certainly makes the case for solar PV much more favourable given that PV systems have a lifetime of at least 25 years.

Although solar HW and solar PV systems currently do not appear to be cost-effective at current fuel prices, they do offer the potential to save a considerable amount of carbon emissions from the UK housing stock. Studies from extrapolation of savings from the Oxford DECoRuM model (and validated with studies by BRE and DCLG) show that, deploying solar hot water and solar PV systems in existing UK dwellings could potentially save about 3.2 million tonnes of carbon (MtC) per year. This equates to almost 7.8% reduction over the domestic sector emissions in 2004 (see Table 5.6) at a capital cost of about £100 billion to £180 billion. Importantly, it is only by integrating solar systems with energy efficiency measures and low carbon systems that we can achieve reductions above 60% from the UK housing stock to meet our national targets. Even ongoing studies by DCLG indicate that under current existing stock conditions and with currently known technologies, to achieve a 60% reduction would require the application of solar energy technologies (DCLG, 2006).

*Table 5.6* **Potential of reducing carbon emissions from the UK housing stock**

| | | |
|---|---|---|
| **UK domestic sector carbon emissions in 2004:** | **41.7 MtC** | |
| **Savings from:** | | |
| Energy Efficiency measures: | 18.3 MtC | 43.9% reductions |
| Low carbon systems (heat pumps and micro-CHP): | 5.7 MtC | 13.7% reductions |
| | *Sub total* | *57.6% reductions* |
| To achieve >60% reductions: | | |
| Solar energy systems (HW and PV): | 3.2 MtC | 7.8% reductions |
| | TOTAL | **65.4% reductions** |
| **Additional capital cost for solar systems (HW and PV)** | **£100bn–£180bn** | |

*Source*: Gupta, 2005a.

## Solar cities

There is a large and growing '*solar city*' movement. This term is defined as being an urban area with active programmes to increase the number of solar systems installed on buildings to:

- increase the energy generated from renewable sources
- to reduce carbon dioxide emissions from those buildings
- and reduce the dependency of the city as a whole on externally generated supplies of energy.

The *solar city* concept has been promoted through a wide range of very different initiatives. These include the use of including financial and rate-based incentives to increase the short-term economic benefits to individuals of installing systems (www.solartechnologies.com/Incentives.shtml) or legal, guideline and regulation-based strategies promoting solar technologies (www.ren21.net/globalstatusreport/gsr4c.asp).

*Solar City* programmes are typically associated with energy efficiency and other Demand Side Management strategies, urban planning and architectural developments to promote the benefits of solar design. An important aim of many *Solar City* programmes is their 'climate-stabilisation' and 'climate change mitigation'. The core emerging indicator sets for such cities concern the trends in the per-capita emissions related to the targets set for future greenhouse-gas emissions at levels consistent with stabilising future atmospheric carbon-dioxide and other greenhouse gases within an established time frame. For case studies of a number of different solar cities see: www.martinot.info/solarcities.htm. The Solar City movement has been largely driven by through the medium of two International Conferences to date (www.solarcities.org.uk) in Daegu and Oxford and a 2008 conference to be held in Adelaide, Australia (www.solarcitiescongress.com.au/)

## The future

A decade ago when the Oxford Ecohouse was built the first question typically asked about the solar systems was 'yes but how much does it cost?', meaning it's too much. In a century in which we will see the almost complete depletion of our global oil and gas reserves, in which the warming climate threatens to undermine our lifestyles and the stability of political systems around the world, with oil and water wars already coming to the fore, surely this is the wrong question? Surely, in light of the building benefits of solar energy outlined above, the question we should be asking is rather 'can we live comfortable, decent and safe lives in homes that are powered to a greater extent by free, infinitely renewable and clean energy form the sun?' The answer in Oxford is yes, and we can do it with traditional building techniques using materials in a conventional construction market with typical construction crews. In Oxford today the same question is asked frequently 'How much does it cost?' and usually followed now by 'and where can I get one?'.

### References

Beazley, E. and Harverson, M. (2002) *Living with the Desert: working buildings of the Iranian Plateau*. Thailand. ISBN: 9748304647: Orchid Press.

Brown, G.Z. and Dekay, M. (2000) *Sun, Wind and Light: Architectural Design Strategies*. ISBN: 0471348775: John Wiley & Sons Inc.

Capra, Fritjof and Gunter Pauli (eds), Steering business towards sustainability, United Nations University Press (1995). Online version accessible on: http://www.unu.edu/unupress/unupbooks/uu16se/uu16se0a.htm. This contains a well reasoned discussion of the relative metrics of various incentives for the uptake of solar energy.

Chadwick, H.M., Batley-White, S.L. and Fleming, P.D. (2002) The UK planning process and the electricity supply industry – what role for renewables? In: *Creating sustainable urban environments: future forms for city living*. Oxford: Christ Church College.

Crisp, V., Littlefair, P.J., Cooper, I. and McKennan, G. (1998) *Daylighting as a Passive Solar Energy Option: an assessment of its potential in non-domestic Buildings*. IHS BRE. ISBN: 0851252877.

DCLG (2006) *Review of the sustainability of existing buildings: the energy efficiency of dwellings – initial analysis.* London: Department for Communities and Local Government.
Edwards, B. and Turrent, D. (2000) *Sustainable housing: principles and practice.* London: E & FN Spon.
Fthenakis, V.M. Overview of Potential Hazards (2003) *Practical Handbook of Photovoltaics: Fundamentals and Applications,* Markvart T. and Castaner L. (eds), Elsevier, pp. 854–68. See also: http://www.earthscan.co.uk/news/article/mps/uan/505/v/3/sp/.
Gadsden, S.J. (2001) *Managing the urban environment: the solar energy potential for dwellings.* PhD Thesis. Institute of Energy and Sustainable Development. Leicester: De Montfort University.
Galloway, T. (2004) *Solar House: A Guide for the Solar Designer,* Architectural Press. ISBN: 0750658312.
German Solar Energy Society and Ecofys (2005) *Planning and Installing Photovoltaic Systems: A Guide for Installers, Architects and Engineers.* Earthscan Publications Ltd. ISBN: 1844071316.
Green Electricity Marketplace (2004) *Domestic tariffs.* Retrieved on 21 August 2004 from the World Wide Web: http://www.greenelectricity.org/regions/southern.html.
Gupta, R. (2005a) *Investigating the potential for local carbon dioxide emission reductions: developing a GIS-based domestic energy, carbon-counting and carbon-reduction model.* PhD Thesis. Department of Architecture. Oxford: Oxford Brookes University.
Gupta, R. (2005b) Investigating the potential for local carbon dioxide emission reductions: developing a GIS-based domestic energy, carbon-counting and carbon-reduction model. Refereed Technical Paper. In: *2005 Solar World Congress.* Orlando, Florida, USA, 2005b.
Hastings, R. *Solar Low Energy Houses of IEA Task 13* (2005) London: James & James.
Henning, Hans-Martin (ed.) (2003) *Solar-Assisted Air-Conditioning of Buildings: A Handbook for Planners* Austria: Springer-Verlag. ISBN: 3211006478.
H.M. Treasury (2003) *The green book: appraisal and evaluation in Central Government.* London.
IEMA (2000) *Managing climate change: a business guide.* London: Institute of Environmental Management and Assessment.
IEA PVPS (2002) *Basics of PV: environmental considerations.* Retrieved on 31 May 2003 from the World Wide Web: http://www.pv-uk.org.uk/technology/index.html.
Iles, P. and Shorrock, L.D. (2004) *Cost effectiveness calculations for fabric measures.* Prepared for Energy Saving Trust, Watford: Building Research Establishment.
Jardine, C. and Lane, K. (2003) *Photovoltaics in the UK: an introductory guide for new customers.* Oxford: Environmental Change Institute, University of Oxford.
Kachadorian, J. (1997) *The Passive Solar House: Using Solar Design to Heat and Cool Your Home.* Chelsea Green Pub Co. ISBN: 0930031970.
Knowles, R. (2006) *Ritual House: Drawing on Nature's Rhythms for Architecture and Urban Design.* Island Press. ISBN: 1597260509.
Markvart, T. (ed.) (2000) *Solar Electricity* (UNESCO Energy Engineering Learning Package, John Wiley and Sons Ltd. ISBN: 0471988537.
Mazria, E. (1979) *Passive Solar Energy Book.* Rodale P, ISBN: 0878572376.

Muneer, T. (2004) *Solar Radiation and Daylight Models: for the energy-efficient design of buildings*. Butterworth-Heinemann Ltd. ISBN: 0750659742.
ODPM (2004) *Proposals for amending Part L of the Building Regulations and implementing the Energy Performance of Building Directive* A consultation document, London: Office of the Deputy Prime Minister.
Porteous, C. and MacGregor, K. (2005) *Solar Architecture in Cool Climates*. Earthscan Publications Ltd. ISBN: 1844072819.
Roaf, S., Fuentes, M. and Thomas, S. (2001) *Ecohouse: a design guide*, Oxford: Architectural Press.
Roaf, S., Fuentes, M. and Gupta, R. (2003) *Feasibility study report on the Oxford Solar Initiative*, Submitted to the Energy Saving Trust. Oxford: Department of Architecture, Oxford Brookes University.
Roaf, S., Crichton, D. and Nicol, F. (2005) *Adapting buildings and cities to a changing climate*, London: RIBA Publications Ltd.
Schaffer, J. and the Real Goods Staff (ed.) (1996) *Real Goods' Solar Living Sourcebook: The Complete Guide to Renewable Energy Technologies & Sustainable Living*, Chelsea Green Publishing Company. ISBN: 0930031822.
Schlaich, J. and Robinson, M. (1995) *The Solar Chimney: Electricity from the Sun*. Edition Axel Menges. ISBN: 3930698692.
Scoones, A. (2001) *Renewable energy in the built environment*. Bedford: Building Centre Trust.
Shorrock, L.D. and Henderson, G. (1990) *Energy use in buildings and carbon dioxide emissions*. Watford: Building Research Establishment.
Shorrock, L.D., Henderson, J. and Utley, J. (2005) *Reducing carbon emissions from the UK housing stock* Watford: Housing Centre, Building Research Establishment Environment.
Solar Energy Applications Laboratory (2005) *Solar Heating and Cooling of Residential Buildings: Design of Systems*. University Press of the Pacific. ISBN: 1410224589.
Solar Energy International (2004) *Photovoltaics: Design and Installation Manual*. New Society Publishers. ISBN: 0865715203.
Strong, S. and Scheller, W. (1993) *The solar electric house: energy for the environmentally-responsive, energy-independent House*. Chelsea Green Pub Co. ISBN: 0963738321.
Szokolay, S. (2003) *Introduction to Architectural Science: The Basis of Sustainable Design*. Architectural Press. ISBN: 0750658495.
UNEP and UNFCCC (2001) *Climate change information kit*, United Nations environment programme's information unit for conventions, Geneva.
Viljoen, A. (1995) *Low Energy Dwellings*. Msc Thesis. University of East London: School of Architecture.
Watson, T.R., Zinyowera, M.C. and Moss, H.R. (1996) *Technologies, policies and measures for mitigating climate change*, IPCC Technical Paper 1, Geneva.
Watson, J. (2002) *Renewables and CHP deployment to 2020* Tyndall Centre for Climate Change Research.
Watson, J. (2004) Co-provision in sustainable energy systems: the case of micro-generation. *Energy Policy*, 32 (17), pp. 1981–90.
Yannas, S. (1994) Solar energy and housing design: volume 1: principles, objectives, guidelines. London: Architectural Association.

# 6
# Saving Energy – How to Cut Energy Wastage

*Bob Everett*

## Introduction

Most of the previous chapters have concentrated on energy supply. This chapter looks at ways to save energy and focuses on energy in buildings in the UK. In 2005, out of the total UK annual $CO_2$ emissions of almost 153 MtC energy use in buildings contributed 61 MtC, 38 MtC from the domestic sector and another 23 MtC from the services sector (offices, schools, hospitals, etc) (DTI, 2006a). The question is how much of these emissions could be saved.

Energy use is, of course, only the means to provide various *energy services*. These include: the provision of warm homes and working environments, cooked food, adequate illumination and the ability to use electronic equipment for communication, entertainment and writing books like this. We could simply do less of everything and turn the heating down. This is *energy conservation*, but here we focus on *energy efficiency*, achieving the same level of energy services using less energy and a lower environmental cost.

## A little history

Life indoors in the past in the UK must have been pretty cold in winter. Nineteenth and early 20th century buildings had solid walls and single glazing. They had to be well ventilated both to supply the combustion air for the coal fires and to get rid of the fumes from the oil and gas lamps. The basic principle of keeping warm seems to have been to wear lots of clothes, sit as close as possible to the fire during the day and retreat under a thick pile of blankets in bed at night. Building standards improved throughout the 20th century. From the 1930s onwards cavity walls were introduced with an air gap between two separate skins of brick, largely as a method to prevent damp penetration. The coal fire remained the normal mode of heating in UK homes well into the 1960s.

With the introduction of North Sea Gas in the 1970s also came gas fired central heating. The proportion of the housing stock with central heating rose from 31% in 1970 to 92% in 2003 (DTI, 2004). With it also came a presumption that houses should be fully heated to an acceptable comfort temperature. Concerns about death rates of the very young and the elderly have given rise to the concept of 'fuel poverty'. A household is said to be in fuel poverty if it needs to spend more than 10% of its income on fuel to maintain a satisfactory heating regime (usually 21°C for the main living area, and 18°C for other occupied rooms). It has been estimated that in 2003 some 2 million households out of the total of 25 million in the UK were in fuel poverty, and had difficulty in keeping their homes warm at an acceptable level of cost. Providing adequate levels of insulation is an obvious solution to this.

Yet even loft insulation was only introduced into the Building Regulations for new UK houses in 1974 and then only to a depth of 25 mm. Since then Building Regulation standards for new buildings have been steadily improved and government campaigns have encouraged householders to install insulation.

However, there is still a large proportion of the existing housing stock that is relatively poorly insulated, particularly in the private rented sector.

The picture for the services sector is not much better. The 1960s saw a fashion for 'curtain wall' office construction where a steel frame was used to provide the structure and the walls were largely made of thin concrete panels and large sheets of single glazing. These offices were hard to heat in winter and often overheated in summer. Fortunately office buildings tend to be regularly refurbished as new occupants come and go, but even so making major improvements to the thermal performance is often ignored.

## UK potential for savings

While Building Regulations are steadily being improved for new buildings, the real problem is what to do about upgrading the existing building stock, its heating systems and the stock of appliances currently being used. For each basic energy uses there are a range of technologies that could be used to reduce demand. These include:

**Space and water heating** – Insulation, better airtightness, ventilation heat recovery, higher efficiency boilers, Combined Heat and Power generation (CHP), heat pumps and better controls

**Lighting** – more efficient lights and better controls
**Electrical appliance energy use** – avoiding standby power, better insulated fridges, more efficient electrical design
**Electric motors** – better design and use of magnetic materials, proper sizing and efficient control systems
**Air conditioning** – avoidance of its use and proper fan and duct sizing where its use is unavoidable

First we will looks at some estimates for the overall potential savings, before going on to briefly describe some of the technologies.

## Domestic sector – the PIU Study

A study by the Performance and Innovation Unit of the Cabinet Office (PIU, 2001) suggested a total energy saving potential of over 1 EJ per year in the existing stock (i.e. approximately 15% of UK delivered energy use). This would result in savings in carbon emissions of nearly 24 MtC. Table 6.1 lists the options studied. It is split into an 'economic potential' and a further 'technical potential'. What is 'economic' has been taken to be those measures having an internal rate of return (IRR) of more than 6%. The IRR is the rate at which it would be worth borrowing money to finance the measure; the higher the rate, the more economic.

This is an impressive list. The total should be treated with some caution because there may be some double counting. For example if a building is insulated first, then the potential for savings with more efficient heating will be reduced. Also, this is only one set of options. More CHP could be used, but that would reduce the potential for the use of condensing gas boilers. Note also that a little bit of 'renewable energy supply' in the shape of solar water heating has been included.

A later study by the Building Research Establishment considered the options above in more detail (Shorrock *et al.*, 2005), producing a similar answer to the PIU, but including 50 PJ per year from floor insulation and a (relatively expensive) potential of over 70 PJ per year from domestic photovoltaic installations.

## Domestic sector – the 40% house project

A study carried out by the Environmental Change Institute at Oxford (Boardman *et al.*, 2005) went further and looked at what would be required to achieve a 60% reduction in $CO_2$ emissions in the UK domestic sector by 2050. Most controversially, rather than embarking

*Table 6.1* **Energy savings and carbon emission reductions from domestic energy efficiency**

| Type of saving | Measure | Energy saving/PJ per year | Carbon saving /MtC per year | Internal rate of Return/ % |
|---|---|---|---|---|
| **Economic Potential** | | | | |
| Insulation | Loft Insulation | 71 | 1.4 | 16% |
| | Cavity Wall insulation | 134 | 2.6 | 32% |
| | Double Glazing | 88 | 1.7 | 19% |
| Heating Systems | Condensing gas boilers | 273 | 5.3 | 27% |
| | Domestic hot water cylinder insulation | 17 | 0.3 | 200% |
| | Heating controls | 21 | 0.4 | 38% |
| | Small-scale CHP | 8 | 0.3 | 19% |
| Electricity | Energy efficient lights | 38 | 1.4 | 50% |
| | Appliances | 80 | 2.9 | 19% |
| **Sub-total Economic Potential** | | **731** | **16.3** | |
| **Percentage of domestic total** | | **37%** | **41%** | |
| **Additional technical potential** | | | | |
| Insulation | Solid wall insulation | 143 | 2.8 | +3% |
| | Draught proofing | 17 | 0.3 | +6% |
| | High performance glazing | 63 | 1.2 | –2% |
| Heating systems | New district heating CHP | 25 | 0.9 | not given |
| | Ground source heat pumps | 17 | 0.7 | 0% |
| | Solar water heating | 84 | 1.6 | –8% |
| **Sub-total** | | **349** | **7.7** | |
| **Total** | | **1092** | **23.8** | |
| **Percentage of domestic total** | | **54%** | **57%** | |

*Source*: PIU (2001).

of a large programme of insulating solid-walled buildings, it suggested that the housing demolition rates would have to increase dramatically. Around 14% of the current stock would need to have been replaced with energy efficient buildings by 2050.

In addition to the measures considered above, it assumed a very large take-up of CHP, with 22% of dwellings having community heating. This is likely to be most used in city centres and provides a way of cutting the energy use in historic buildings. Outside city centres the ECI study assumed that a further 40% of dwellings would use individual domestic micro-CHP. It also assumed high insulation levels and the use of high efficiency triple glazing.

As for electricity use by appliances and lighting, the study only projected a 27% overall fall in national consumption from 1998 levels. While a four-fold reduction in electricity use seems possible in cold appliances and lighting, electricity use by TVs and consumer electronics could well double by 2050.

In order to reach the 60% target it, like the BRE study, included a large take-up of solar thermal water heating and PV panels. Even so, it was necessary to assume that the overall carbon intensity of electricity from the grid would fall by a quarter by 2050 through the use of more renewable energy in electricity generation.

## Services sector

A study carried out for the Royal Commission on Environmental Pollution looked at the services sector (Fisher *et al.*, 1998). It suggested that the economic energy saving potential here was 22%, resulting in emission cuts of 5.1 MtC. Approximately 10% of this potential was through the use of small-scale CHP. It estimated that the technical potential for savings was nearly double that at 39%, with emission savings of 8.2 MtC.

## Policies and changing attitudes to investment

The last 30 years have seen a sea change in attitudes to investment in energy saving in the UK. In the early 1970s the idea that homes should be insulated at all seemed revolutionary. By the early 1990s energy saving was a deemed a 'good idea' but it should be tempered by 'commercial rates of return', i.e. 15% discount rates or 5-year payback times. In Table 6.1, dating from 2001, the discount rate has fallen to 6%. Since then, the introduction of 'carbon trading' has added a carbon price (typically £70 per tonne of C, or £20 per tonne of $CO_2$) into the calculations. In the recent review of the Building Regulations (ODPM, 2004) the 6% discount rate was further dropped to 3.5%. The 2006 Building Regulations, which specify for the first time U-values for refurbishment work on existing dwellings, suggest using a 15 year payback time to assess what is 'reasonable'. Even more recently, the cosy vision of a secure future of cheap and freely available gas has been challenged and the calculations may need to be done yet again.

The overall result is that energy efficiency measures that seemed hopelessly uneconomic in the 1990s, and even marginally economic by 2000 are being considered as fully economic in 2006. If a 'global warming panic' sets in and carbon prices rise (a figure of £140 per tonne C is contemplated) or there is an equivalent 'gas security of supply panic' then a whole new range of measures will become economic.

Currently the process of improving the existing building stock is being pushed forward by the EU Energy Performance of Buildings Directive (EPBD) (EU, 2002). This follows in the footsteps of previous EU directives for appliances (described later).

The EPBD requires Member States to put in place by 2007:

a) a methodology of calculation of the integrated energy performance of buildings
b) a procedure for setting minimum energy standards in new and existing buildings
c) the Energy Certification of large buildings, together with regular inspection of their heating and cooling installations.

The UK has already had a working methodology for calculating the energy performance of houses for some years, in the form of BREDEM, the Building Research Establishment Domestic Energy Model. This now, in its incarnation as the Standard Assessment Procedure (SAP), forms the core of the 2006 Building Regulations for dwellings. The wording of these for new buildings is perhaps a little obscure, but essentially say 'take the 2002 regulations (which were very detailed) and produce a building with 20% less $CO_2$ emissions – you are free to use the SAP to do the sums'. This allows future Regulations to be tightened up by simply amending the 20% improvement to a higher figure.

The Energy Certification of existing houses and flats is being introduced from June 2007 with the Home Information Packs (HIP). This requires an energy survey (by a qualified inspector) to give an Energy Performance Certificate, giving a 'SAP rating', an estimate of the $CO_2$ emissions and an energy label rating on a scale from A-G like fridges. On this scale the average UK house would get an 'E' rating and one built to the 2006 Building Regulations a 'B' rating. However, it will not require that homes are brought up to a specific standard (which is the next logical step).

Its counterpart for large buildings, the Standard Building Energy Model or SBEM is not quite so far advanced. There is no doubt that one of the aims of the EPBD that all public buildings should have their 'energy label' on display will eventually materialise. This is the first step before the inevitable task of mandatory 'upgrading to a standard' can take place.

## Saving heat energy – the technologies

Low temperature heat energy is used in buildings for space heating and water heating. This totalled some 2.3 EJ in the UK in 2003 (DTI, 2004). There are two basic ways of reducing this demand:

(i) cutting the losses of the building fabric (walls, floor, roof and windows) by the use of insulation, and the ventilation losses with better airtightness and possibly using ventilation heat recovery
(ii) installing a more efficient heating system or using CHP. This is also a good way of saving on energy use for hot water provision (the next step being to install solar water heating).

## Better insulation and airtightness

### Insulation

Given a history of cheap and abundant coal followed by one of cheap natural gas, it is not surprising that the UK building stock is poorly insulated. Yet the materials to rectify the situation are hardly rocket science. Rockwool and fibreglass insulation are available in the DIY sheds. The higher grade plastic foam insulation materials such as polyisocyanurate and polyurethane are now appearing in builder's merchants yards. 'Green' insulation materials such as cellulose fibre and sheep's wool insulation are still only available from specialist suppliers.

It is important to understand that 'insulating a building' to modern standards requires a considerable thickness of even the best-performing materials. There is no magic 'insulating paint' that can be applied in some conveniently thin layer.

The heat loss of a building element is specified by its U-value. An uninsulated roof or a solid brick wall can have a U-value of around 2 $W/m^2K$. Target U-values for modern construction or refurbishment work can be a tenth of this value. For example the current recommended thickness of 250 mm of wool insulation should reduce the heat loss from its uninsulated state by more than ten-fold, giving a U-value of 0.16 $W/m^2K$.

The walls of buildings can be insulated. In existing buildings with cavity walls insulation can be injected into the gap. Typically this will improve the U-value from about 1.5 $W/m^2K$ to about 0.6 $W/m^2K$. This can only be done if the building is not exposed to driving rain, since the main function of the cavity is to stop damp penetration through the wall.

Solid walls (and cavity walls) in existing buildings can be 'dry lined'. A layer of insulation can be put on the inside faced with plasterboard. This is fairly simple to do. The U-values that can be achieved are mainly dependent on how much reduction in the interior room sizes can be tolerated. For example using 100 mm of rock wool will improve the U-value to better than 0.45 $W/m^2K$ (EST, 2004). Alternatively they can be external insulated, usually with a thick layer of plastic foam which is then either covered with a layer of cement render or a special cladding layer. This is commonly being done in the refurbishment of tower blocks of flats. Although this is relatively expensive, it is possible to achieve good U-values of better than 0.3 $W/m^2K$ with 100 mm of foam insulation and better than 0.2 $W/m^2K$ with 150 mm.

In new construction insulation batts can be incorporated into the cavities of brick walls with aerated concrete blockwork used to build the inner leaf. A wall U-value of 0.35 $W/m^2K$ is suggested by the 2002 UK Building Regulations. In practice, this can easily be bettered. The cavity can be made as wide as necessary (150 mm to 300 mm) to incorporate more insulation and, if necessary, to retain an air gap to prevent damp penetrating the wall. Timber frame construction can also use considerable thicknesses of insulation as necessary. Two hundred millimetres or more of wall insulation is commonly used in Scandinavia and Germany.

## Insulated glazing

Double glazing was only made mandatory for new houses in England and Wales in 2002. This was extended in 2005 to cover replacement windows for existing houses. The regulations also specified a maximum U-value of 2.0 $W/m^2K$ for windows with wood or PVC-U frames, or 2.2 W/m K for aluminium or steel frames. These values are what could be achieved with 'hard coat' low emissivity (low-e) coatings to reduce the heat loss across the gap between the two panes. This involves the use of a thin layer of tin oxide on the glass. However, an improved emissivity is now commercially available using a 'soft coat' process

*Table 6.2* **Indicative U-values for windows with wood or PVC-U frames**

| Glazing type | W/m²K |
|---|---|
| Single glazing | 4.8 |
| Double glazing (normal glass, air filled) | 2.7 |
| Double glazing (hard coat low-e, emissivity = 0.15, air filled) | 2.0 |
| Double glazing (hard coat low-e, emissivity = 0.2, argon filled) | 2.0 |
| Double glazing (soft coat low-e, emissivity = 0.05, argon filled) | 1.7 |
| Triple glazing (soft coat low-e, emissivity = 0.05, argon filled) | 1.3 |

*Source*: BRE (2005) SAP 2005.

using very thin layers of optically transparent silver. This coating is, however, more delicate and it is difficult to produce a low-e toughened glass for use in such applications as glazed doors.

The heat loss across the gap between the panes can be reduced by filling it with the heavy gas argon, or even better the rarer and more expensive gas krypton. Table 6.2 gives some indicative U-values for different glazing options.

It has been estimated that in 2004, 43% of UK households had some form of double glazing on 80% or more of their windows (BRE, 2006). That, of course, means that there is still a potential for improvement for the other 57% of households.

## Air-tightness

Proper air-tightness is the key to minimising the heat loss due to ventilation. In existing housing this means using draught-stripping, replacing leaky windows and closing off unused chimneys. The latter may be difficult since it is often necessary to maintain a small air-flow through them to remove any moisture penetrating into them. It means paying careful attention to blocking off all the unwanted air leakage paths while maintaining the essential ones.

In new construction attention to detail is really important. It is all too easy to leave air gaps around windows and where pipes penetrate walls. Sheet plastic vapour barriers are often built into walls, especially in timber-framed construction. For really good airtightness these vapour barriers must be taped together where they join, so that they cover the whole building envelope. This is quite a skilled job. The latest Buildings Regulations for new construction require sample pressure tests to check the air leakage.

## Mechanical ventilation with heat recovery

UK buildings an particularly houses are traditionally naturally ventilated. However, once the fabric heat losses of a building have been tackled with thick insulation, the major remaining major heat loss will be that from ventilation. Mechanical ventilation with heat recovery (MVHR) involves pumping air through ducts to and from the various rooms of a building. The warm outgoing air is allowed to preheat the cold incoming air by passing both air streams through a heat exchanger.

MVHR is a mixed blessing. On the one hand it gives controllable ventilation adjustable to every room. On the other it requires complex ductwork and air pumping which can consume considerable amounts of electricity. MVHR systems are available for domestic applications but it is essential that they are installed in buildings that are really airtight to start with. Otherwise any attempt to pump air around the system may just increase the flow of air through unwanted air infiltration paths.

## Future insulation standards

A typical poorly insulated UK house of the 1970s might have a net space heating demand of 13,000 kWh, while one built to the 2002 Building Regulation might achieve a net space heating demand of 4,000 kWh, i.e. a reduction of 70%. If a house were sufficiently well insulated – superinsulated – solar gains and heat from the occupants and appliances should be sufficient to keep the internal spaces warm in all except the coldest weather. A 'space heating system' as such might be very small, or even completely unnecessary.

This has been the philosophy of the German 'Passive House' programme, which has looked at not only new buildings, but also retrofitting existing ones to superinsulated standards. This may require using fabric U-values of approximately 0.1 $W/m^2K$ (Feist *et al.*, 2005). A demonstration project at Ludvigshaven in south west Germany, the '3-litre house' gave an estate of apartment blocks a thorough thermal modernisation, including:

- at least 200 mm of foam insulation in the roof and walls
- triple glazed windows with low-e glass
- mechanical ventilation with heat recovery

Monitoring showed that the space heating energy use was reduced by a factor of seven to only 30 kWh (or 3 litres of heating oil) per year per square metre of floor area (equivalent to 3,000 kWh per year for a 100 $m^2$ flat) (Luwoge, 2006).

The German climate is, of course, colder than that of the UK. Here, the Association of Environmentally Conscious Builders (AECB) has proposed a 'Gold Standard' for new houses with roof U-values of 0.1 W/$m^2$K and 0.14 W/$m^2$K for the walls and floor, together with use of the best triple-glazed windows (AECB, 2005).

## Improving heating efficiency

Over the past 50 years the UK has been transformed from a country where the majority of homes were heated with coal fires to one where central heating is almost universal. In 2004, 92% were centrally heated and this proportion is expected to continue to rise. Gas is the dominant central heating fuel. Renewable energy and energy from waste meet less than 1% of UK domestic heating energy demand (BRE, 2006).

This growth of central heating use has been a mixed blessing for overall energy demand. On the one hand it is estimated that the average space heating efficiency has risen from about 50% in 1970 to over 70% in 2004. On the other hand homes are now better heated and estimated average internal temperatures have risen from about 12°C to 18°C over the same period, increasing the potential heat losses.

## Gas and oil-fired boilers

The efficiencies of gas and oil-fired boilers have improved enormously over the past 30 years, from about 65% to over 90%. This has been brought about by the adoption of:

**Electronic spark ignition** – replacing permanently burning pilot lights which could consume 10% of the total gas.

**Balanced Flues** – here the burner and heat exchanger is totally sealed off from the interior air of the building and the combustion air is blown through the boiler using an electric fan. Although this design does require a certain consumption of electricity it saves energy by allowing the building to be made more airtight.

**Condensing Gas Boilers** – the main component of natural gas is methane. When this burns it produces a large amount of water vapour. In a non-condensing boiler this is lost with the flue gases. In a condensing boiler the heat exchanger is made sufficiently large to enable the water vapour to be condensed out, recovering its latent heat of vaporisation. This can increase the amount of heat extracted from the gas by up to 11% and increase the overall boiler efficiency to 90% or more. Condensing boilers are available for use with heating oil, but the potential efficiency improvements are slightly lower, about 6%, because of the lower proportion of hydrogen in the fuel.

The efficiency values of different designs can be found in the government's SEDBUK (Seasonal Efficiency of a Domestic Boiler in the UK) rating database (available at http://www.boilers.org.uk).

In order to simplify things for purchasers boilers have also been given an Energy Rating Label. Those with a SEDBUK of more than 90% are 'A' rated, those with a SEDBUK figure of 85%–90% are 'B' rated, while those with a SEDBUK of below 70% are 'G' rated. From 2005 the UK Building Regulations only allowed 'A' and 'B' rated boilers to be installed for most applications.

## Combined heat and power generation (CHP)

An alternative to using individual boilers in buildings is Combined Heat and Power generation. This uses the waste heat from power stations, of many possible sizes. The potential for this is enormous. In 2003, 32% of UK primary energy was 'lost in conversion and delivery and energy industry own use'. The bulk of this, over 2 EJ, was lost as low grade waste heat from power stations. This figure should be compared with the 2.3 EJ of low temperature heat used for space and water heating.

In Denmark, the equivalent loss figure is only 22%. The country has adopted a national policy of widespread use of CHP with district or community heating, i.e. distributing the heat through large insulated pipes under the streets. Over a half of their electricity comes from CHP and 12% of their primary energy goes into district heating. Although most of this comes from large power plants, they also have many small-scale CHP plants, often in rural areas fired by waste straw (Danish Energy Authority, 2005).

In the UK in 2005 CHP only supplied about 8% of the country's electricity. The bulk of the 5.8 gigawatts (GWe) of generation capacity was large gas or steam turbine plant in industry (DTI, 2006b).

## Small-scale CHP in the UK

In buildings there was about 350 MWe of CHP plant spread over 1,100 separate installations. Over 95% of these were spark ignition gas engines. Typically units range in size from about 50 kWe up to 1 MWe, with electrical efficiencies of from 25% up to 40%, larger units being more efficient. Overall fuel efficiency can be 80% or more. Prime users are buildings with large heat and electricity loads such as hospitals, hotels and leisure centres with swimming pools.

This is a technology that has been heavily promoted in the UK in the past. However it has been hampered by an electricity market which requires that any electricity exported must be precisely predicted hours in advance. Thus there is a tendency for CHP units currently being installed to be sized only to meet the local building electricity demand without exporting any electricity. This rules out taking advantage of the economies of scale in using large machines.

## Stirling engine domestic Micro-CHP

Over the past ten years, small CHP units of approximately 1 kW electrical output have been developed, intended as replacement domestic gas boilers. These use a small Stirling engine to drive a generator. The Stirling engine is an external combustion engine, widely used in the 19th and early 20th centuries as an alternative to the steam engine for powering small machinery. Its main advantage over the internal combustion engine is that it can be almost silent. These are packaged as a condensing gas boiler which also generates electricity. They are primarily designed to meet the heat load of a house but not to generate the full electricity demand as well. Unlike larger small-scale CHP units they are designed to export electricity which the installer will purchase.

It is suggested that if large numbers of these were to be installed the total amount of electricity produced at any given time would be a reasonable approximation to the average domestic electricity load. Their installation would thus have a beneficial effect on the grid in reducing both average and peak transmission loads from conventional power stations.

The electrical generation efficiency for these commercial units is not high, a figure of 10–12% being quoted for the 4-cylinder Whispergen. However, prototype machines have been built with efficiencies of 30% or more.

The potential for this technology is enormous. A report by the Society of British Gas Industries (SBGI, 2003) suggests that these could be potentially be retrofitted in 14 million UK homes, with an estimated annual $CO_2$ emission saving of 1.5 tonnes per year per unit.

## Fuel cell CHP

Small-scale CHP units using fuel cells are now becoming commercially available and are widely used in the USA and Japan. Fuel cells generate electricity from hydrogen and oxygen. In practice, for the present, the hydrogen is likely to be produced by a 'reformer' from natural gas while the oxygen comes from the air.

In 2001, a 200 kWe phosphoric acid fuel cell (PAFC) was installed at a leisure centre in Woking as part of a scheme incorporating a further 1 MWe of gas engine CHP and 9 kWp of photovoltaics. The individual fuel cell elements are stacked together to produce approximately 400 volts DC, coupled to the grid *via* an AC inverter. Hot water from the fuel cell is used not just to provide heating but also to run an absorption chiller unit for cooling for the leisure centre. Performance monitoring showed that the fuel cell operated with approximately the same electrical generation efficiency, 37%, as a comparable gas engine CHP unit (DTI, 2005).

Small prototype domestic fuel cell CHP units are also undergoing trials. The Proton Exchange Membrane Fuel Cell (PEMFC) may be the most suitable for these because it can be started up in a matter of minutes whereas other types may best be used continuously. These prototype systems are being produced with an electrical output of 1–1.5 kW, an electrical generation efficiency of over 35% and an overall fuel efficiency of over 80% (SBGI, 2003).

At present both the capital costs and maintenance costs are higher than for gas engines, but these could fall in the future. Most of the maintenance costs of the Woking unit were for the natural gas reformer rather than the fuel cell itself. Fuel cell CHP has two main advantages over engine-driven CHP: the units are almost silent, with no moving parts and they have high electrical efficiencies which in the longer term could reach 50%, i.e. competitive with large CCGT power stations (IEA, 2005).

## Using waste heat from large power stations

Despite being widely used in Denmark and Germany, this is an option hardly used at all in the UK, since the trend has been to build large coal-fired power stations well away from major built up areas. Although many gas-fired power stations have also been built on green field sites, there are several close to major urban centres. For example the 1 GWe Barking CCGT station is less than 20 km from central London and pumps hundreds of MWs of waste heat into the Thames.

There is obviously enormous potential for using waste heat from such large power stations, but it comes at a cost of reduced electricity generation efficiency. The trade-off between heat and electricity is known as the Z factor. A Z factor of 6 means a loss of one unit of electricity for every 6 units of heat produced and is typical for low pressure steam taken from a steam turbine power station. It has been suggested that a Z factor of 10 could be achieved with a purpose built CCGT CHP station (PB Power, 2005). Heat from CHP plant can be seen has having a very low marginal energy and $CO_2$ emission cost. It can therefore be argued that its use is better than embarking on expensive retrofit insulation for existing buildings.

## Which form of CHP is best?

There are many possibilities for CHP. Generally, the larger the CHP unit, the higher the electricity generation efficiency, the lower the capital costs per kW of capacity and the lower the maintenance costs per kWh of electricity produced.

A recent International Energy Agency report (IEA, 2005) modelled the heating needs of a European city of about 250,000 inhabitants (based on Leicester). It compared different forms of CHP with a base case of using individual condensing gas boilers and electricity from new high-efficiency CCGT plant. It concluded that the least cost solution and the one with the largest $CO_2$ emission savings was city-wide CHP using heat and electricity from a CCGT power station. However, this leaves many buildings in suburban areas for which individual small CHP units may well be the best solution. However, if small fuel cell CHP units can be produced with high efficiencies and low maintenance costs then these could become serious challengers to city-wide CHP.

A further possibility is to include heat storage in a CHP system. This would then allow the CHP unit much more flexibility of operation without losing too much in the way of overall fuel efficiency. Thus CHP units could be become a flexible complementary electricity generation technology to variable renewable sources such as wind, PV and tidal power. Work on this is currently being carried out at Birmingham

University under the EU 'DESIRE' programme (Dissemination strategy on energy balancing for the large-scale integration of renewable energy).

## Making CHP happen

According to a Building Research Establishment study there is an economic potential for 18.3 GW of CHP in UK domestic, commercial and public buildings. This would allow 5.5 million homes to use CHP i.e. about a quarter of the housing stock (BRE, 2002). The study was based on using reciprocating gas engines, but comments that once large areas were connected there could be economies of scale by switching to using larger CCGT plant.

The government has set a target of 10 GWe of CHP capacity by 2010, yet it is extremely unlikely that this will be met. Although things have improved since 1988 when there was under 2 GWe of CHP capacity, progress is hampered by the nature of the UK electricity market. This is just not friendly to small electricity exporters. Most of the increase in UK CHP capacity has been in the form of large gas and steam turbine plant in industry.

Between 1999 and 2004 the amount of installed capacity in buildings only increased by 38 MWe (DTI 2000; DTI 2005b), which can hardly be regarded as a success story. There is a argument for some form of guaranteed 'feed-in tariff' to encourage small-scale CHP. Indeed such a tariff was put in place in the 1983 Energy Act, only to be removed when the electricity industry was privatised in 1989.

City-wide CHP with district heating is also strangely ignored despite having an enormous potential and a long history of studies. It was reviewed in the late 1970s by the Department of Energy under its chairman, Sir Walter Marshall (later Lord Marshall of Goring). This group published two reports, Energy Paper 20 (DoE, 1977) and Energy Paper 35 (DoE, 1979). The reports concluded that '*When oil and natural gas are no longer available for heating, the potential for CHP/DH could possibly be in the region of 30 per cent of the existing domestic, commercial and institutional heat load in the UK*'. They pointed out that about half of the national heat load was associated with Greater London and 80% with the five largest conurbations, viz Greater London, West Midlands, Greater Manchester, Merseyside and Glasgow. It urged that heat load mapping of urban areas was carried out which, indeed, has now been done.

The reports stressed the need to set up district heating schemes and asked for a 'Heat Board' to take on the national responsibility for CHP and district heating. The UK government consistently refused to set up any kind of coordinating heat authority. Lord Marshall warned that this would be a 'recipe for indefinite delay' and so far has been proved right. This in contrast to the vigorous action taken in Denmark at the same time.

It is also worth pointing out that Lord Marshall was equally enthusiastic about nuclear power. While the 1970s Energy Papers were prepared to consider nuclear CHP, this is a subject about which little is heard today. Yet it makes little sense to consider building new nuclear power stations only to throw two-thirds of their heat into the sea.

## Ground source heat pumps, a solution for buildings beyond the gas grid?

What about those buildings that are aren't connected to the gas grid and have no local CHP plant? Are they doomed to use increasingly expensive heating oil or electric resistance heating? One solution is to use a ground source electric heat pump (GSHP). Its basic principle is the same as that of the refrigerator. Heat is taken from a heat exchanger buried the ground, raised in temperature using mechanical work delivered by an electric compressor and then delivered to the interior of a building. There it can be distributed *via* a conventional central heating system. The ground source heat exchanger is a set of pipes either in a tube drilled vertically downwards or in a horizontal trench at least 1 metre deep.

A key parameter of a heat pump is its coefficient of performance (COP), the ratio of heat delivered to the electrical energy supplied. In a monitored UK installation the COP averaged nearly 3.2 over the year (EST, 2000), i.e. every kWh of electricity produce more than three kWh of heat. In practice the COP is dependent on the temperature difference through which the heat pump has to operate; the larger the difference, the worse the performance. This is why ground source heat exchangers are favoured over air source ones (as used for air conditioning plant). Indeed the UK electricity industry abandoned a programme of air source heat pump research in the 1970s when it was found that their performance in the coldest weather was very poor. As a result they didn't cut peak winter electricity demand. Using a heat exchanger buried in the ground gives it a more stable temperature environment and helps tide the system over really cold weather.

Although gas engine driven heat pumps have been produced the GSHP is essentially an electric heating technology. Its attraction is that it reverses the primary energy losses at the power station. If it takes 2.5 kWh of heat to generate 1 kWh of electricity, this can then be turned back into 3 kWh or more of delivered heat energy. Thus in terms of the overall ratio of primary to delivered energy it can be more efficient than using gas boilers. At present the running costs and $CO_2$ emissions for GHSPs in the UK are estimated to be only slightly lower than for gas-fired central heating, but considerably better than for oil-fired central heating or electric resistance heating (EST, 2005).

## Saving electricity

### Electrical appliances

Cutting electrical demand in buildings requires tackling a whole range of uses. There has been a steady tide of new applications for electricity ever since mains electricity was introduced in the 1890s, from cookers and electric irons through to TVs and mobile phones. It does not seem sufficient for each household to own just one of each; by 2004, UK households owned on average 2.4 televisions each (EST, 2006).

Not only is the ownership of appliances per household increasing, so is the number of households. The UK population is likely to increase from almost 61 million in 2006 to around 67 million by 2050, but also the average household size is falling. It is currently about 2.5 persons per household, but this is likely to fall to 2.1 persons by 2050. The total number of households could thus rise from about 25 million today to nearly 32 million by 2050 (Boardman *et al.*, 2005). Into every new household goes a new set of appliances. In 2004 electricity use by domestic appliances and lighting amounted to 89 TWh, about a quarter of total national demand.

## Standby power

Perhaps the most disturbing wastage of electricity is that from millions of small devices that are not fully switched off but are left on 'standby'. The standby electricity used just by televisions in the UK left translates into $CO_2$ emissions of about 0.15 MtC per year. Add in all the other forms of equipment, audio equipment, video or DVD players, computers, photocopiers, etc., which revert to standby mode when not in use and this emission figure rises to 1.2 MtC per year (House of Lords Science & Technology Committee, 2005).

Devices such as televisions have been designed for 'convenience'. To allow the TV to be turned on by a remote control from an armchair, the power supply is left running permanently, just to power a small infrared sensor and a relay. The situation has not improved with the development of digital television. This requires an integrated receiver decoder (IRD) or set-top box (STB) specifically designed to be left on permanently to allow 'off-peak' downloading of TV programmes for 'view on demand'.

Another problem arises with 'mains adapters' and battery chargers. Electronic devices are sold world-wide. In order to simplify compliance with safety regulations about high voltages in appliances they are often designed to be run from low voltage supplies. A separate adaptor is then be supplied to provide the required voltage (6 V or 12 V) from the local mains supply. Thus although the actual electronic device may be switched on and off, the mains adaptor or charger usually remains plugged in and permanently switched on. The situation is compounded by the use of low quality transformers which still draw current through the primary coil although none is being taken from the low voltage secondary coil. The result is a continual dissipation of heat.

Since these devices often made in the Far East, this is a global problem. The solution is better design and enforced international standards. The standby power of new designs is being limited in a new 2005 European Energy-using Products (EUPs) Directive. The regulation for standby power says quite simply 'As a general principle, the energy consumption of EuPs in stand-by or off-mode should be reduced to the minimum necessary for their proper functioning' (EU, 2005). A long-term target for a maximum standby power is 1 watt. This Directive should become law in the UK in 2007.

## Televisions and consumer electronics

In use a typical CRT television of today consumes about 100 watts and a large-screen plasma TV may use 500 watts. These and other electronic devices represent an increasing use of electricity in the home. In the office, the increased use of electronic equipment is a source of excess heat which may promote the installation of air conditioning. Yet the electronics industry is capable of great feats of energy efficiency if pressed. The laptop computer is carefully designed to eke out the absolute maximum period of operation from the minimum battery capacity. Liquid crystal displays (LCDs) are universally used in them.

They can use only 10% of the electricity of an equivalent CRT display. Although LCD displays are likely to compete with plasma screens for the 'widescreen' market, screen sizes and consequence power ratings are likely to increase in the future. The '40% House' study suggests that UK domestic electricity use in consumer electronics could double by 2050 to over 20 TWh per year.

## Refrigerators

In contrast, these might be described as having the seeds of a success story. In 2004 domestic refrigeration in the UK used an estimated 15 TWh of electricity per year, almost 4.5% of total demand (DTI, 2006). This figure is actually down from a peak of 17.5 TWh in 1999 due to the improved energy efficiency of new fridges on sale. However even these figures have to be compared with the 2 TWh used for refrigeration by the food processing industry which presumably handles exactly the same food and stores it for longer periods, but it does so in physically larger and more energy efficient cold stores.

Until the 1980s there was little interest in the UK in producing 'energy efficient' fridges. Yet the potential for saving was enormous. Following suggestions that the energy consumption of cold appliances could be reduced by up to 50%, in 1994 the European Commission introduced an energy labelling Directive. This uses an 'A'–'G' rating which reflects the consumption in kWh per litre of net volume. An 'A' rated appliance uses about a half of the electricity of an 'E' rated one (Schiellerup, 2002). In 1999 a further Directive set minimum standards phasing out the sale of fridges with a rating below 'C' (except for chest freezers where it could be 'E'). The original labelling system was not very ambitious and at present (2006), many manufacturers offer not just A rated appliances but also 'A+' and even 'A++' ones that are more efficient than the rating system envisaged.

Current 'A' rated fridge designs use about 25 mm of insulation thickness while freezers use about 50 mm. Improved efficiency can always be achieved with thicker insulation. However, the external dimensions are fixed in practice by the standard units of width and height of fitted kitchens, so improved efficiency will usually mean less internal storage space.

The way forward may be through the use of *vacuum insulation panels* (VIPs). One manufacturing approach is to take a plastic foam or an 'aerogel' made from silica, wrap it in a thick airtight plastic cover and pump any air or gases out. As long as the foam has sufficient strength not to collapse under the atmospheric pressure, very good thermal conductivities can be achieved representing a 4–5 fold improvement over the best non-evacuated plastic foam insulation. There are still many problems to be solved to maintain this performance over a long period of time since any pinholes in the outer layer could destroy the vacuum. Although VIPs are now commercially available, they are expensive and there are likely to be trade-offs of production cost against life expectancy (Glacier Bay, undated).

If VIPs can be developed and the manufacturers' reluctance to trade storage volume for insulation thickness overcome, then the fridges of 2050 could each use under 100 kWh per year. UK domestic total electricity use by 'cold' appliances could have fallen by a factor of five from its peak 1999 level to only 3.5 TWh per year (Boardman *et al.*, 2005).

## Energy efficient lighting

In 2004 lighting in buildings accounted for about 15% of UK electricity use. The domestic sector used approximately 18 TWh and the services sector a further 37 TWh, of which over a third was used in retail premises.

There are many types of lighting available. Table 6.3 below lists the main ones, together with their characteristics and *luminous efficacies*, the amount of light emitted by a lamp in lumens per watt (lm/W) of electricity used. In general for a given type, lamp efficacy increases with wattage. Very high efficacies are available for street lighting, but most of the applications in buildings require small lamps. Despite the availability of compact fluorescent lamps, standard low wattage incandescent lamps and small tungsten halogen lamps remain widely sold. Those in domestic use can have efficacies of under 20 lumens per watt.

Light emitting diodes (LEDs) are relative newcomers to the list. Although familiar as small indicator lights, they have now been developed in a range of colours, including now a bluish white. One attraction is their potentially long life, tens of thousands of hours, compared to 1,000–2,000 hours for an incandescent lamp or 8,000 hours for a compact fluorescent lamp. Although they are currently mass-produced as small lights of under 1 watt output, these can be built up into modules that can compete with small incandescent lamps. At present efficacies in commercial lamps are around 25 lm/W, but a performance of over 100 lm/W has been demonstrated. In the longer term these have the potential to produce dramatic cuts in domestic lighting consumption. It has been suggested that by 2050 this could have fallen by

***Table 6.3*** **Lighting efficacies of modern electric lamps**

| **Lamp type** | **Details** | **Lighting efficacy lumens per watt** |
|---|---|---|
| General Service Lamp | The common incandescent light bulb | 9–19 |
| Tungsten Halogen | Miniature incandescent lamps, often run on a 12 volt supply | 17–27 |
| Light Emitting Diodes | Available in a range of colours and bluish white | 25–100+? |
| Fluorescent high pressure mercury discharge lamp | Bluish white light, often using for shop lighting | 40–60 |
| Metal Halide high pressure mercury discharge lamp | Bluish white – used for street lighting | 75–95 |
| Compact fluorescent lamp | miniature fluorescent replacement for GLS incandescent | 70–75 |
| Tri-phosphor tubular fluorescent | office and shop lighting | 80–100 |
| Sodium discharge lamp | orange light, used for street lighting | 75–200 |

a factor of four from current levels to around 4 TWh per year (Boardman *et al.*, 2005).

## Efficient ventilation and cooling

This is a slightly controversial subject. Traditionally UK buildings have been naturally ventilated. However, as described above, cutting winter ventilation heat loss may require the construction of airtight buildings and the use of mechanical ventilation with heat recovery. This in turn will require the use of electricity to pump air around the building.

Also UK buildings have not traditionally required cooling. However, climate change brings with it the possibility of higher summer temperatures which may prove life-threatening. The 2003 heat wave in France with peak daytime temperatures of over 35°C is estimated to have caused 14,800 excess deaths, particularly amongst the elderly (Pirard *et al.*, 2005).

In 2004 'cooling and ventilation' in the UK services sector used about 9 TWh of electricity, 4% of its total energy use. Most of this was used in fan power rather than refrigeration. But is this energy use really necessary? Analysis of the energy consumption of UK office buildings has shown a wide variation. A typical 'prestige' airconditioned office can use nearly three times as much energy per square metre of floor area as a typical naturally ventilated one (Action Energy, 2003). In part this variation is due to different office functions. Even so, for a given type of office, those using 'good practice' design used 40% less energy than a 'typical' example.

Mechanical ventilation requires the use of electrically driven fans and ductwork to circulate air within a building. In a domestic installation, the circulation fan may only be rated at 50 watts and the ductwork will only a few tens of metres in total length. In the system of a large multi-storey office or hospital there may be a kilometre or more or steel ductwork usually hidden above a false ceiling. The fans may require large electric induction motors of 100 kW or more, which may run continuously day and night 365 days a year.

If mechanical ventilation is necessary, there are ways to minimise the electricity consumption. For example, it is important to make sure that the ducts are large enough. A narrow duct may save on metalwork but will result in more fan power to pump the air.

Fan motors offer a good opportunity for reducing electricity use, by:

- using a high efficiency electric motor
- avoiding oversizing of the motor
- using a variable speed drive motor drive

The techniques listed above also have enormous potential in the industrial sector, where there are an estimated 10 million motors installed with a total capacity of 70 GW and their use makes up two thirds of the total electricity demand (ETSU, 1998).

## Alternative methods of cooling buildings

There are a number of ways to cool a building avoiding some of the disadvantages of electric air conditioning. For example, as used in the Woking scheme above an *absorption chiller* (using principles similar to those used in gas-powered refrigerators) can use waste heat from a CHP plant instead of electricity. Alternatively, the building can be cooled by night ventilation. This has been done in the Elizabeth Fry Building at the University of East Anglia in Norwich. It has used the Swedish Termodeck system of thick concrete floor slabs, which are hollow and can be cooled by circulating outside air through them at night when air temperatures are lower. Its monitored gas and electricity consumption was less than half that of a 'good practice' air conditioned office building in the UK. (Standeven *et al.*, 1998). If electric air conditioning is unavoidable, then its performance can be improved by using ground source heat pumps rather than air source ones, or 'off-peak' air conditioning can be used, making ice at night for cooling during the day.

## Smart meters

Current electricity (and gas) metering is little changed from the late 19th century. Meters are read maybe one or twice a year. Surely we can do better than this with 'smart meters' in a information age? There are several clear needs:

**Provision of clear general energy use information** – if customers had clear instantaneous feedback on their electricity use, then they might be able to take action to reduce their total consumption. Without it, they will remain ignorant.

**Peak demand tariffs and information for all customers** – in the UK, the peak demand for electricity occurs between about 17:00 and 20:00 on winter weekdays. This is a time when offices and industry are shutting down for the day and domestic electricity consumers are arriving home, cooking meals and watching TV. Although part of this peak demand can be met from pumped storage plant, much of it has to be met from generation plant that will sit idle for most of the rest of the year. Although commercial electricity meters have a 'peak demand' facility to discourage use at this time, this is not present in domestic meters or tariffs. Experiments in California have shown an average peak demand reduction of 13% where customers were given warning of 'super-peak' prices (45 p/kWh). This is a matter of 'shifting demand' and delaying such activities as washing clothes to a time of lower electricity prices.

**Import-export metering for microgeneration** – the development of distributed generation in the form of domestic micro-CHP, PV panels and wind turbines brings with it the need for two-way metering and billing for both imported and exported electricity.

**Remote scheduling of domestic micro-CHP and appliance use** – on-peak/off-peak meters in the UK are already remotely radio controlled (*via* the Radio 4 Long Wave transmitters). It is only a short step to foresee a fully 'informated grid', where there is a parallel two-way flow of information about electricity use and generation. Thus if a load were deemed to be relatively unimportant, it would not be supplied until the electricity price had fallen to an appropriate level or a certain amount of time had elapsed. Domestic micro-CHP plant could also be remotely controlled to generate on demand rather than just meeting the heat needs of their particular house. It has also been suggested that an 'informated grid' could give rise to whole new decentralised markets for electricity (Awerbuch, 2004).

At present the capital cost of 'smart meters' appears to be a stumbling block. There is a perverse logic to this – if buildings can be made to consume little energy, then it does not make much sense in paying for expensive meters to measure it. On the other hand, if these meters allow the flexible use of limited energy supplies, it may be money well worth spending.

## Conclusions

This chapter has described a wealth of opportunities for saving energy in buildings. The total economic and technical energy saving potential in the domestic and services sectors according to the PIU and RCEP studies is approximately 1.4 EJ per year or roughly 50% of current use. The bulk of this potential lies in heat energy savings through the large-scale application of very ordinary technologies: thicker insulation, double glazing and condensing gas boilers.

To achieve carbon emissions cuts of 60% from the UK building stock will need further steps. These might include digging up the streets to distribute waste heat from power stations to most city centre buildings, or the development and deployment of millions of domestic micro-CHP units. It may even require the demolition of large numbers of 'energy-unfit' houses.

The technical potential for cutting electricity use is enormous, yet it does seem an uphill struggle against a tide of new devices which are designed to be 'attractive' and 'convenient' rather than 'energy efficient'.

Most importantly, we live in a culture where the provision of 'energy services'; buildings that are warm in winter and adequately lit, has been seen largely as a matter of supplying cheap gas and electricity. The needs of facing up to climate change and diminishing North Sea gas supplies seem to be changing attitudes to energy saving. As a result the Building

Regulations have been tightened up quite dramatically in the last decade, and they may be tightened further in the next one. However, we are left with an inadequately trained building workforce and a trail of out-of-date attitudes to energy saving which urgently need updating.

## References

Action Energy (2003), *Energy Consumption Guide 19: Energy use in offices* (downloadable from http://www.carbontrust.org.uk – accessed 8th July 2006).

AECB (2005), *Gold and Silver Energy Performance Standards for Buildings*, Association of Environmentally Conscious Builders (downloadable from www.aecb.net – accessed 29th July 2006).

Awerbuch, S. (2004), *Restructuring our Electricity Networks to Promote Decarbonisation*, Tyndall Centre Working Paper 49 (downloadable from http://www.tyndall.ac.uk/ publications/pub_list_2004.shtml).

Boardman *et al.* (2005), 40% House, Environmental Change Institute, Oxford.

Building Research Establishment (2002), *The UK Potential for Community Heating with Combined Heat & Power*, downloadable from http://www.est.org.uk/housing-buildings/communityenergy/ (look under community energy publications).

BRE (2005), *SAP 2005 – The Government's Standard Assessment Procedure for Energy Rating of Dwellings*, Building Research Establishment, Watford downloadable from projects.bre.co.uk/SAP2005 (accessed 8th May 2006).

BRE (2006), *Domestic Energy Fact File 2006*, Building Research Establishment (downloadable from http://www.bre.co.uk accessed 3/06/2006).

Danish Energy Authority (2005), *Heat Supply in Denmark*, downloadable from http://ens.netboghandel.dk/english/ (look under publications).

DoE (1977), *District Heating Combined with Electricity Generation in the United Kingdom – Energy Paper 20*, Department of Energy, HMSO, London.

DoE (1979), *Combined Heat and Electrical Power Generation in the United Kingdom – Energy Paper 35*, Department of Energy, HMSO, London.

DTI (2000), *Digest of UK Energy Statistics (DUKES)*, Department of Trade and Industry, London.

DTI (2003 – tables updated 2004), *Energy Consumption in the UK* (downloadable from http://www.dti.gov.uk/energy accessed 27/11/06).

DTI (2004), *Energy Consumption in the UK* (updated from 2003 edition) Department of Trade and Industry, London.

DTI (2005a), *Woking Park PAFC CHP Monitoring* (downloadable from http://www.dti.gov.uk/energy).

DTI (2005b), *Digest of UK Energy Statistics (DUKES).*

DTI (2006a), *Energy Sector Indicators 2006*, downloadable from www.dti.gov.uk/energy – accessed 29th June 2006.

DTI (2006b), *Energy in Brief*, downloadable from www.dti.gov.uk/energy.

EST (2000), *Heat pumps in the UK – a monitoring report*, GIR72, Energy Saving Trust (downloadable from http://www.est.org.uk).

EST (2004), *Energy Efficient Refurbishment of existing housing*, CE83, Energy Saving Trust (downloadable from http:// www.est.org.uk).

EST (2005), *Best Practice in new housing – a practical guide*, CE95, Energy Saving Trust (downloadable from http://www.est.org.uk).

EST (2006), *Rise of the machines*, Energy Saving Trust, London (downloadable from http://www.est.org.uk – accessed 7th July 2006).
EU (2002), *Directive 2002/91/EC on the energy performance of buildings* (downloadable from http://www.managenergy.net/products/R210.htm – accessed 18th July 2006).
EU, (2005), *Directive 2005/32/EC establishing a framework for the setting of ecodesign requirements for energy using products* (downloadable from http://ec.europa.eu/enterprise/eco_design/directive_2005_32.pdf, accessed 23rd June 2006).
ETSU (1998), *Energy Savings with Motors and Drives*, Good Practice Guide 002 (downloadable from http://www.thecarbontrust.co.uk, accessed 1/05/2006).
Feist, W. *et al.* (2005), *Climate Neutral Passive House Estate in Hannover-Kronsberg: Construction and Measurement Results*, Darmstadt, Germany (downloadable from http://www.passivhaustagung.de/englisch/texte/PEP-Info1_Passive_Houses_Kronsberg.pdf – accessed 30th July 2006).
Fisher *et al.* (1998), *Prospects for Energy Saving and Reducing Demand for Energy in the UK*, Royal Commission on Environmental Pollution, London.
Glacier Bay (undated), *Vacuum Insulation Panels (VIPs) Principles, Performance and Lifespan*, Glacier Bay Inc, California (downloadable from http://www.glacier-bay.com/vacpanelinfo.asp, accessed 24th June 2006).
House of Lords Science & Technology Committee (2005), *Energy Efficiency, 2nd Report of Session 2005–6*, downloadable from http://www.parliament.uk (accessed 1/05/2006).
IEA (2005), *A Comparison of distributed CHP/DH with large-scale CHP/DH*, International Energy Agency.
Luwoge (2006), *The 3-Liter House* (downloadable from www.luwoge.de).
ODPM (2004), *Proposals for amending Part L of the Building Regulations and Implementing the Energy Performance of Buildings Directive: a consultation document*, Office of the Deputy Prime Minister, London.
Pirard *et al.* (2005), *Summary of the mortality impact assessment of the 2003 heat wave in France, Eurosurveillance*, Vol. 10, downloadable from http://www.eurosurveillance.org – accessed 5th July 2006.
PIU (2001), *Energy Efficiency Strategy*, Performance and Innovation Unit, UK Cabinet Office (downloadable from http://www.strategy.gov.uk/downloads/files/PIUc.pdf – accessed July 1st 2006).
PB Power (2005), *The Supply of Heat from Barking Power Station to the Borough of Barking and Dagenham*, PB Power, London.
SBGI (2003), *MicroCHP – delivering a low carbon future*, Society of British Gas Engineers.
Schiellerup, P. (2002), *An examination of the effectiveness of the EU minimum standard on cold appliances: the British case*, Environmental Change Institute, Oxford, downloadable from http://www.eci.ox.ac.uk/lowercf/pdfdownloads/ECEEE01_PS.pdf.
Shorrock *et al.* (2005), *Reducing carbon emissions from the UK housing stock*, Building Research Establishment, Watford (downloadable from www.bre.co.uk).
Standeven, M. *et al.* (1998), *PROBE 14: Elizabeth Fry Building*, Building Services Journal April pp. 37–42 (downloadable from http://www.usablebuildings.co.uk/Probe/FRY/FRYAPR98.PDF – accessed 8th July 2006).

# 7

# The Limits to Energy Efficiency: Time to Beat the Rebound Effect

*Horace Herring*

While the previous chapter has presented technical solutions to reducing energy use, this chapter first looks at the historic failure of this approach to reduce consumption and then at how we might do better. The simple message is that achieving reductions in energy use, and consequent greenhouse gas emissions, is not simply a matter of technology. What also matters is how we use the technology and what our expectations of it are. This can be illustrated by an example.

If you look at the box containing a Compact Fluorescent Lamp (CFL) bulb the only thing it guarantees is its wattage, its light output and its lifetime. It may label itself as 'energy saving' but it does not give you any indication of how much energy you will save from it. For in reality and despite the hype, a CFL, like most other energy efficiency devices, are power, not energy, saving devices.

The energy use of an appliance is a combination of two factors: power and time on. The manufacturer may control the power rating, but you, the consumer, control the time it is on. The only appliance where the annual energy consumption can be predicted with any certainty is for those that are continuously connected, like fridges and freezers. But here again there can be wide variations in use for identical appliances, depending upon the conditions under which they operate and how they are used. A fridge placed next to a cooker or boiler, will use more than one in a garage. A freezer whose door is left open or is not defrosted regularly will use more. So saving energy is not just about technology but about our habits, attitudes and lifestyle, including our attitudes to time. To summarise, energy efficient technologies may be power saving but not necessary energy saving, since consumers often change the way they operate efficient devices. i.e. they may use them longer, more intensively, or buy more of them. In particular, the energy savings resulting from energy efficiency equipment depend crucially on our time management, on whether we are time-poor or time-rich. For time scarcity drives consumption decisions and therefore influences energy consumption in many ways (Dimitropoulos, 2006).

## Time scarcity

If we are to improve on current approaches to energy efficiency then the time dimension has to be taken into account more generally – since it also interacts with attitudes to costs and effects consumer purchasing decisions. For example time influences whether people actually get around to installing energy efficiency devices in the first place. For people are saturated with ads every day to buy this or do that. Even if they agree with the energy efficiency message – and few dispute the merits of cutting costs and saving money – their time is limited and they have to set priorities for action. It is basically the 'hassle factor' – how much effort is involved in getting a quote for loft insulation by phone, or buying an energy efficient light bulb? People subject their effort to a crude cost-benefit analysis: what will I save for this effort, is it worth the hassle! Academics refer to this hassle as 'transaction costs' and often these present a formidable obstacle to achieving action.

Is five minutes on the phone worth saving £50 on the motor insurance, is 15 minutes on the web worth saving £5 on a book, is an hour spent shopping in the high street worth saving £10 a year on an energy efficient fridge? The easier it is made to save the more people will respond, hence the development of selling services, like insurance, travel and banking, on-line. It avoids the chore of visiting shops or phoning around. It also, most crucially, save consumers time. Thus it should be no surprise that consumers fail to take advantage of all the offers to save money through energy efficiency. Firstly they may not believe the offer, secondly they may think the offer is not relevant to them, thirdly they just haven't got the time to make the effort, and finally there are more fun things to do in life that worry about saving £10, or even £50, a year.

The government's response to consumer indifference to saving energy is to mutter about their 'economic irrationality' and advocate the need to 'educate' them. But what is the best way this could be done? Should there be education campaigns to persuade them of the error of their ways (preaching), or should they be forced to change their minds through regulation and higher prices (the stick) or bribed through incentives and subsidies (the carrot)? Should we be more efficient through moral persuasion (it is good for us and good for the planet) or through economic self-interest (we can save money and spend it on luxuries)?

Likely responses to these various types of intervention will depend to some extent on how consumers perceive the impact on their lifestyles, and how willing they are to change. A range of possible approaches, involving the conscious adoption on different lifestyles, will be discussed later – many on them call for a switch to a less material but time-richer way of life. These lifestyle issues are the subject of a new five-year research programme at the University of Surrey (RESOLVE, 2006).

However, there are some types of consumer behaviour that seem to be almost involuntary, at least within the existing context of energy use. In this situation, attempts to achieve cuts in energy use may be thwarted, and this may explain the failure so far to achieve reductions.

## The rebound effect

The UK government has a goal of a 60% cut in $CO_2$ emissions by 2050, and expects improved energy efficiency to lead to a reduction in national energy use. But is this feasible? Despite decades of improvements in energy efficiency – and factor 4 and even factor 10 improvements have been made over the last century – national energy consumption continues to rise. There are many reasons put forward for this, such as population and economic growth, technical change and higher standards of living. However these factors are expected, and even encouraged, for the future, so why would energy consumption fall in the future? What will make the future different from the past?

One explanation put forward for the failure of energy consumption to fall despite efficiency improvements is the 'rebound effect' (or take-back effect). This is the term used to describe the effect on consumption of an efficiency improvement, which by lowering the cost of producing or operating a product or service encourages people to use more of it. For instance when we replace a 75 W incandescent bulb with an 18 W compact fluorescent bulb (CFL) – a reduction in (wattage) power of about 75% – we could expect over time a 75% energy saving. However it seldom is. Consumers realise that the light now cost less per hour to run, and so are often less concerned about switching it off, indeed they may intentionally leave it on all night. Thus they 'take back' some of the energy savings in the form of higher levels of energy service (more hours of light). This is particularly the case where the past level of energy services, such as heating, were considered inadequate.

The energy savings from efficiency improvements, such as increased levels of insulation, may then be spent on much higher heating standards – the consumer benefits by getting a warmer home for the same or lower cost than previously. In addition consumers may use new technologies in unexpected ways that increase energy use, rather than decrease it. This has occurred with conservatories (see Box 7.1).

The rebound effect is the focus of a long-running dispute within energy economics but until recently was confined to the general obscurity of academic journals. Then in 2005 the House of Lords Science and Technology Committee gave a sympathetic hearing to the economist Len Brookes, the leading UK critic of the proposition that increased energy efficiency would lead to a reduction in national energy use (a view termed the Khazzoom-Brookes postulate). In its report the Committee said:

> The Government's proposition that improvements in energy efficiency can lead to significant reductions in energy demand and hence in greenhouse gas emissions remains the subject of debate among economists. The 'Khazzoom-Brookes postulate', while not

*Box 7.1* **The conservatory paradox**

Originally, conservatories were thought likely to reduce household energy consumption. In theory, this 'buffer space' should passively collect solar energy, pre-heating the house and reducing the need for additional central heating. However, this theory relied on the conservatory being used during warmer months only, rather than all year round as part of the main house. A survey done in 1993 found the reality is quite different. Of 1800 respondents, 91% said they heat their conservatories and 50% of those said they do so regularly. A recent update of this work shows trends towards year-round conservatory use and the installation of heating and now also cooling systems is increasing. The irony is that such spaces lose heat at ten times the rate of conventional insulated rooms, even when double-glazed. Thus conservatories substantially increase energy usage by some households rather than delivering the 10% reduction in overall household energy usage originally predicted.

*Source*: Oreszczyn, T. (1993) The energy duality of conservatory use. *Proceedings of the 3rd European Conference on Architecture* 17–21.

> proven, offers at least a plausible explanation of why in recent years improvements in 'energy intensity' at the macroeconomic level have stubbornly refused to be translated into reductions in overall energy demand.
>
> (HoL Science and Technology Committee, 2005, Paragraph 7)

Following this criticism, the government asked the UK Energy Research Centre (UKERC) to undertake a systematic review entitled 'The Evidence for a Rebound Effect from Improved Energy Efficiency' and a final report is expected in mid 2007 (UKERC 2007). In conjunction with this work, the UK Department of Environment, Food & Rural Affairs (Defra) commissioned research to model the macroeconomic impacts for the UK of the rebound effect (Defra, 2006).

## Types of rebound

There are three main types of rebound effect. First, the *direct rebound effect,* is the increased use of energy services caused by the reduction in their price due to greater efficiency. This works exactly as would the reduction in price of any commodity and has immediate effects. Second, the *indirect effect,* is caused by increased expenditure on other goods and services, due to the financial savings arising from use of energy efficient products. For instance, some people might spend the savings from insulating their home on a foreign holiday! Third, the *economy wide rebound effects* is the effect of efficiency changes on the direction and pace of technical change and innovation in the economy (Herring, 2005).

Energy efficiency improvements by both consumers and producers initiate a chain of effects that have repercussions throughout the economy. Energy efficiency in production reduces the cost of final outputs and makes consumption goods cheaper. This increases real income for consumers and thereby increases consumer spending. At the same time, energy efficiency improvements lower production costs, increase producers' margins and increase production output. Higher margins may trigger a price war and encourage competitor firms to adopt energy efficiency improvements (spillovers). All these processes encourage economic growth and, all else being equal, increase aggregate energy demand. This acts to offset the original reduction in demand brought about by the efficiency improvements.

Most probably the greatest effect (in the long term) of lower costs of energy services is on the direction and pace of technical change and innovation in the economy. For new goods and services are created to take account of the possibility of lower energy costs. For instance, innovators and manufacturers aware that the cost of lighting with CFLs has fallen by 75%, will devise new lamps for new lighting uses, like security or flood lighting, or for lighting previously unlit areas, such as the garden or patio. The market for lighting thus increases and consumption has increased by 50% in the last three decades. This is largely due to the lifestyle attractions of 'mood' and multiple point lighting, and as we purchase these new lamps, total energy consumption soon outstrips our original savings. We may have energy efficient lights but we have far more of them, and leave them on longer. Similarly with domestic appliances: they are getting more efficient but we own far more of them, so total energy consumption rises. In fact between 1972 and 2002, electricity consumed by them has doubled (EST, 2006).

Successive UK governments have made improving insulation levels in housing a top priority, and while insulation levels have risen and appliances become more efficient, total energy use in households has risen too. This was partly due to an increase in the number of households but also due to higher heating standards (more widespread central heating systems and higher indoor temperatures) made feasible by insulation and boiler efficiency improvements. Energy is also so cheap, by historic standards, that we can afford to heat the outside air, with patio heaters!

Much of the evidence on the magnitude of the (direct) rebound effect comes from US transportation studies where there is good statistical data on travel miles per vehicle and petrol consumption The results indicate that the number of vehicle miles travelled will increase (or rebound) by between 10% and 30% as a result in improvement in fuel efficiency. Similar results are obtained in the US for domestic energy services (Greening *et al.*, 2000). So the direct rebound effect is generally less than 50%, although it can be over 100% for those suffering from 'fuel poverty', as energy efficiency improvements give them the ability to heat their homes to a level which previously they could not afford (Hong *et al.*, 2006). However it is the impact of energy efficiency on long-run economic growth that is the hardest to examine both theoretically and empirically, and is thus the area where the greatest dispute lies. More broadly, what is the role of increased efficiency of resource use, including energy and time, in stimulating economic growth, increased consumption and greater affluence? This is an issue which is beyond the scope of this chapter, but which is explored by the UKERC Rebound project (UKERC, 2007).

An illustration of the complexities and linkages between efficiency and consumption is given by the following history of lighting.

## Seven centuries of lighting

For lighting it is possible to track changes in efficiency and consumption over seven centuries, as has been done in a most fascinating study by Roger Fouquet and Peter Pearson (2006) for the UK. There they trace the evolution of demand for lighting as technology of lighting progress through medieval candles, 18th century oil lamps, 19th century gas lights and finally 20th century electric lamps. Every time a new technology is introduced efficiency is improved and consumption increases dramatically. Our modern electric lights are 700 times more efficient than the oil lamps of 1,800, and our consumption – measured in lumens-hours per capita – is over 6,500 times greater. Even in the era of the electric light, over the past 50 years, there has been a doubling of efficiency but a fourfold increase in per consumption (See Table 7.1 below).

Over the last two centuries the cost per capita for lighting per hour has only doubled – the price (0.03%) times the consumption (6641) – while GDP has gone up by a factor of 15. So lighting has become much more affordable and as we highly value illumination as an energy service it is not surprising that we consume so much more light.

## Greater efficiency or less consumption

Historically there has been an unending race between energy efficiency and economic growth. Although the UK uses energy far more efficiently than a century ago – energy intensity has fallen by a factor of 3 since 1880 – national energy consumption has still grown. Despite economies becoming more energy efficient, energy consumption still rises with economic growth (thought the linkage is not 1:1). Even with our

***Table 7.1*** **The economics of electric lighting**
Index 1,800=1

| | Price of lighting fuel | Lighting efficacy | Price of light per lumen | Consumption – lumen-hours per capita | Real GDP per capita |
|---|---|---|---|---|---|
| 1,800 | 100% | 1 | 100% | 1 | 1 |
| 1,850 | 40% | 4 | 26.8% | 4 | 1 |
| 1,900 | 26% | 7 | 4.2% | 86 | 3 |
| 1,950 | 40% | 331 | 0.15% | 1,544 | 4 |
| 2,000 | 18% | 714 | 0.03% | 6,641 | 15 |

*Source*: Fouquet and Pearson 2006. Seven Centuries of Energy Service: Table 3.

current emphasis on energy efficiency our energy use still increases. For instance in the UK over the last 35 years energy efficiency (as expressed by energy intensity – a rough proxy) doubled – a Factor 2 improvement. However GDP more than doubled, so total energy consumption rose by a about 15%. Thus at current rates of efficiency improvement, it is perfectly feasible for there to be a Factor 4 improvement in the next century. But as the Royal Commission on Environmental Protection commented:

> There will continue to be very large gains in energy and resource efficiency but on current trends we find no reason to believe that these improvements can counteract the tendency for energy consumption to grow. Even if energy consumed per unit of output were reduced by three-quarters or Factor Four, half a century of economic growth at 3% a year (slightly less than the global trend for the past quarter century) would more than quadruple output, leaving overall energy consumption unchanged.
>
> RCEP, 2000

Some authors, however, argue that modern economies can, and indeed are, restructuring towards low energy intensity service industries; a process they term 'dematerialisation'. The end result they say, is that we will be able to 'decouple' economic from energy growth. However all economic growth involves the use of some energy; it is hard to think of any (economic) activity that does not involve some energy use, perhaps the nearest would be leisure activities that take place in the open air, like classic concerts, sports or car boot sales! One innovation proposed is that the 'ICT' or 'digital' economy (otherwise known as the 'new', 'weightless' or 'knowledge' economy), will bring forth 'dematerialisation'. However after the 'dotcom' euphoria of the early 2000s there is no hard evidence that this innovation will automatically or necessarily be good for the environment. Propaganda here, like that for the earlier 'paperless' office, has outstripped evidence, and some research show that ICT can increase consumption, particularly if it stimulates personal travel or freight traffic (Rejeski, 2003). For instance buying books on the internet can mean they are air-freighted to you!

The internet can also increase the desire for travel and meetings, and has certainly aided the development of low-cost airlines which has greatly stimulated tourism. Its effect is similar to previous revolutions in communications (newspapers, the telegraph, the telephone, radio, TV) over the last 150 years. The knowledge that these technologies bring stimulates transport growth though opening up new commercial possibilities and the desire to travel to new places.

## Sustainable energy

One way consumers can reduce their carbon emissions is to use renewable energy sources. You can generate your own with micro-power units, for example using roof-top micro wind and PV solar. This currently appeals to only a minority as it is a relatively costly and time consuming project, but if energy proves stay high enthusiasm for this approach may spread. However it is a lot simpler and quicker to sign up to a green power tariff scheme which only takes a few minutes by phone or on the web. Around three million consumers in the EU have signed up to such schemes whereby the power they use is matched by their energy suppliers using power from a commercial renewable energy source.

One way to tackle the rebound effect in households would be to use the monetary saving from any efficiency investment to pay for the extra cost of buying in renewable energy. Another approach might to be for retailers to offer a package, with discounts on green power tariff schemes, when consumer bought efficient appliances. Thus the savings from energy efficiency could subsidise renewable energy. The act of switching over the green power may also provide an incentive for consumers to reconsider other energy and environment related aspects of their lifestyles; and a recent report has indicated that this is a common experience (Hub, 2005).

On its own however, just switching to renewables for domestic power is not enough. We also use energy outside the home and renewables do have an environmental impact: there are after all limits to how much of the country we may want to cover with wind farms or biomass plantations. However, alongside a more general move to a sustainable approach to consumption in this and other sectors, it is certainly part of the solution.

## Sustainable consumption

One future policy promoted by the UK government that hopes to break linkage between economic and energy growth is the goal of sustainable development, which can be split into two areas: one on production, and the other on consumption. Sustainable consumption is composed of three elements:

- consuming more efficiently,
- consuming differently, or quite
- simply consuming less

Consuming more efficiently relies on the idea of improving 'resource productivity' or 'eco-efficiency'. The success of sustainable consumption (as a concept) so far has rested on the belief that it is easy to consume more efficiently or achieve 'Factor 4', by using energy and resources more productively. However while resources may be used more efficiently, it does not mean that less are used. The second plank of sustainable consumption is to consume services rather than goods, based on the idea that services consume less resources than products. Again the use of services rather than products does not necessarily mean lower use of resources, particularly if it involves higher standards of services or the extensive use of transport, or other infrastructure such as telecommunications networks, to deliver the service (e.g. home deliveries of internet shopping or take away meals).

However, adopting more efficient (or greener) products without reducing the growth in consumption will not make a large difference in the long term. Research shows that while adopting green consumption patterns can produce a reduction in energy in the short term of 10% to 20%, these reductions are soon outpaced due to rising levels of consumption, caused by even modest growth (1% to 2% per annum) in income (Alfredsson, 2004).

The most contentious aspect of sustainable consumption is about consuming less, as it requires we examine our lifestyle and question our consumption patterns. Even if we are prepared to acknowledge that our 'quality of life' is not really about material consumption, it is very hard for most of us to give up our possessions, reduce our shopping, or to cut down our consumption of resources generally. We can always rationalise and give good practical reasons for our consumption: we need a car because there is no bus to work and it is so much more convenient; we need to fly abroad for holidays because it is cheaper and sunnier than staying at home; we all want to do our bit for the environment but only if it costs little, saves us money and is convenient.

As a report from the Royal Society remarked: '*Unsurprisingly, the concept of sustainable consumption is not popular with governments. It wins few votes and provides an implied threat to competitiveness, employment and profitability. Instead citizens are encouraged to spend, spend and spend*' (Heap & Kent, 2000: 1). So is spending and consuming less an option? Can people change their behaviour towards less material and energy consumption?

## Changing behaviour

Governments already have well established regulatory approaches to reduce energy consumption, like energy taxes, building regulations, and even rationing in emergencies. But can they do anything about changing people's behaviour? Are people willing to reduce their consumption? The answer seems to be 'yes' if it is a group activity and if the 'sacrifice' of foregone consumption is equally shared (SCR 2006) Changing your behaviour is much easier if you are part of a group, which provides encouragement and support, and there are such self-help groups promoting energy efficiency.

For instance Global Action Plan, an environmental charity, adopt a collective, community-based strategy, called the EcoTeam approach. This is a group of six to ten people who might be neighbours, members of the same religious organisation or of some interest group or club. They meet once a month and their eight months programme is based on a workbook which addresses six areas in turn: waste, gas, electricity, water, transport and consumption. The emphasis is on the household rather than the individual, so that EcoTeam members work with other household members to change behaviour.

One long-established route to changing behaviour is through communal living. These vary from the small urban commune to the large rural community, and include co-housing schemes and eco-villages. Such communities have been put forward as an important component of a low consumption lifestyle, with the belief that the shared use of facilities will reduce use of energy and resources. However results from a preliminary study reveals that it is not sharing of resources that reduces consumption but the mutual reinforcement of attitudes towards a low-consumption lifestyle (Herring, 2003).

## Low-energy housing

The easiest way to live a low-energy lifestyle is to live in low-energy house. These have efficiency standards far beyond current building regulations and are generally powered by renewable energy sources. One well known scheme is the Hockerton Housing Project near Nottingham, which consists of five houses that generate their own energy, harvest their own water and recycle waste materials. The houses are amongst the most energy efficient dwellings in Europe and their electricity is generated from two wind turbines and solar cells. Low-energy housing ideas are being incorporated into more mainstream developments, such as at BedZed, in South London, and Gallions Park in London's Dockland, part of a new 'urban village' of over 1,500 new homes, shops and a school, arranged around a central 'ecological corridor'. Gallions Park is an affordable housing scheme of 39, two-, three- and four-bedroom houses, with its developers saying that its aim is to demonstrate that sustainable living isn't something new and cutting-edge, needing a huge effort and an immense budget.

Living in such low-energy housing can lead to significant energy and carbon savings, especially if the technical systems work as planned. Frequently, however innovative renewable systems initially perform below expectations, and it may be the residents' lifestyle rather than the technology that leads to the early savings (Slavin 2006). Particularly significant in such projects are the fostering of community spirit which lead to the success of shared facilities, such as car clubs.

## Car clubs

In the car club concept a fleet of cars is available to individuals who pay according to distance driven. The idea is that there are environmental benefits when cars are shared. In comparison to privately owned cars, club cars are serviced more regularly by the operator which increases their efficiency, the cars are updated quicker to more efficient models since they are used more intensively, and their customers tend to drive less, using public transport or bicycles for short trips.

About 125,000 people use car clubs in Europe, and another 60,000 in North America. In Europe they are most widespread in Switzerland, the Netherlands and in parts of Germany, and are just starting in Britain. One club in Britain is WhizzGo which has six city locations. There is a £25 joining fee, an hourly rate of £5 or £35 for the day, 30 miles free each day, and 20p per mile thereafter – including petrol. Members can prebook by phone or internet to pick up a Citroen C3 at its specified reserved parking bay, and swipe a smart card to trigger the onboard computer to unlock it. In Leeds members qualify for 15% discounts on local transport season tickets, and the aim is to tie in their car club usage with more public transport journeys, while halving their travel costs compared to owning a car. In London a similar scheme from the Streetcar offers its customers cars from over a 100 pick up points for £4.95 an hour, including fuel for up to 30 miles. After that it costs 19 p a mile.

Joining a car club results in significant transport energy savings. In the UK two-thirds of former car owners saw their mileage fall by an average of around 25% and they increased their use of non-car transport modes by 40% (Ledbury, 2004).

## Downsizing

However living in a low-energy house or car sharing is no guarantee of living a low-energy lifestyle: the savings on heating or transport bills might be spent on flying abroad for exotic holidays! For it is your total expenditure (or consumption) – largely determined by your income – that dictates your environment impact or ecological footprint. Poor people and countries have less of a footprint than rich people and countries. Thus one solution, advocated by deep greens like Ted Trainer and Richard Douthwaite, is 'downshifting' or adopting 'voluntary simplicity' (McDonald *et al.*, 2006). That is voluntarily reducing consumption of materials and energy through a lifestyle approach based on earning and consuming less. This might involve living in small self-sufficient communities, which use renewable energy, grow their own food, produce and trade their own goods and services. The ideal is to have a better 'quality of life' (often in a rural area) rather than higher material consumption.

Downshifting, often involving an escape to the countryside, rests on the theory that 'quality of life' is becoming increasingly divorced from economic growth, which does not take account of the cost of crime, pollution and environmental degradation. It finds expression in alternative indicators of economic welfare such as the 'Measure of Domestic Progress' used by the New Economics Foundation. However 'downshifting' is not solely motivated by environmental concerns. Generally it is motivated by the desire to trade highly-paid jobs involving long hours of work and commuting, for less stressful and shorter working hours, and to spend more time on family life. Downshifting need not be long term or permanent – it can be seen as a short-term career break involving part-time working due to family commitments not ecological concerns. Some will return to their high paid careers later when they no longer want (or need) to trade lower income for more leisure.

## Slow movement

Another way to change your lifestyle is by adopting the ideas of the 'slow movement' which is particularly appealing to time-poor urban professionals who are seeking a better work-life balance (Honore, 2004). It is opposed to the cult of speed and efficiency, and is best known for the Slow Food movement founded in Italy and dedicated to leisurely consumption of fine cooking using top quality ingredients, besides preserving traditional ways of food production and retailing, such as organic farming, local varieties and farmers markets. From food the movement has expanded to Slow Cities, dedicated to giving control to pedestrians and small shops rather than cars and chain stores. As Honore concludes: 'What the world needs, and the Slow movements offers, is a middle path, a recipe for marrying la dolce vita with the dynamism of the information age. The secret is balance: instead of doing everything faster, do everything at the right speed' (p. 239). And what exactly is the 'right speed'? All he can answer is 'Sometimes fast. Sometimes slow. Sometimes somewhere in between.'

So can society consume less by working less, and by having more leisure time? Historically we have chosen more income rather than more leisure and downshifting is likely to remain, as it always was, a minority activity. English countryside living or the mythical Italian lifestyle of the 'la dolce vita' can be one vision for a green future but overall the emphasis, if we are to achieve sustainability, will have to be on cities where the majority of people live (Hounsham, 2006). Visions for these will have to concentrate on new ways of living, working and travelling and these are the themes for many science fiction and imaginative stories.

## Personal carbon rationing

The most radical measures currently being discussed is the personal carbon allowance, which would cover all household and personal transport energy use including air travel. There would be an equal carbon allowance for all adults, and trading would be an integral part of a carbon allowance scheme. Frugal people who saved energy, used renewables, and traveled less could sell their surplus to those profligate ones who lived in inefficient homes or travelled a lot. In a speech in July 2006 Mr. Miliband, the UK environment secretary said: 'We should look more radically at the option of tradable personal carbon allowances. Imagine a world where carbon becomes a new currency. We all carry carbon points on our bank cards in the same way as we carry pounds. We pay for electricity, gas and fuel not just with pounds but carbon points' (Adam, 2006).

The carbon allowances would gradually decrease over time, in response both to the need to reduce global emissions and to allow for the expected rise in national population. This would have severe consequences for international travel, since for example just one return flight from London to Athens could exceed the whole personal carbon allowance for the year in 2030 (Hillman and Fawcett, 2004). Carbon allowances may seem unrealistic in our age of consumer choice and freedom, but it is being taken increasing seriously by politicians and policy makers. It may be a necessary measure in the future when there is a need for drastic and rapid cuts in global carbon emissions.

## What is the 'good life'

Technological progress has bought us far more energy efficient goods and services but not lower consumption. We have also chosen greater material consumption over more leisure time, which has lead to us being time-poor. Thus it is not surprising that we are willing to trade energy consumption for some more spare time for consumption. Our lifestyles are limited not by money or energy but by time. Many of us want more time to shop or to travel. Energy efficiency is welcomed if it saves us money, and doubly so if it saves us time.

Ultimately the key questions are ethical not technical, cultural rather than economic. What is the 'good life'? Can we consume more goods and services (for a higher quality of life) but use less materials and energy? Can a low(er) energy lifestyle be made desirable by moral suasion or cultural example? At what stage do we say we have enough energy? We are being urged to accept a 60% reductions in our $CO_2$ emissions. This will undoubtedly require us to use less energy. As the energy analyst Vaclav Smil so passionately argues (Smil 2003: 338):

> Such reductions would call for nothing more than a return to levels that prevailed just a decade, or no more than a generation, ago. How could one even use the term sacrifice in this connection? Did we live so unbearably 10 or 30 years ago that the return to those consumption levels cannot be even publicly contemplated by serious policy makers because they feel, I fear correctly, that the public would find such a suggestion unthinkable and utterly unacceptable?'

The call to return to lower consumption levels asks us to devise a policy of energy 'sufficiency' – that is living well on less energy. Chapter 12 looks at what the options and implications might be for various types of household. This new 'conservation' lifestyle will take time and requires much political consensus and many practical solutions. It will require that we need to delink economic growth from energy consumption. This requires, as we all acknowledge, a step change in the rate of energy efficiency improvement, something we have seldom ever achieved. Energy efficiency is a key tool is this quest to achieve sufficiency but we have to think very clearly about our attitudes to consumption and how we use our time.

## References and further reading

Adam, David (2006). Swipe-card plan to ration consumers' carbon use. *The Guardian* 19 July 2006.

Alfredsson, Eva (2004). 'Green' consumption-no solution for climate change. *Energy* 29/4: 513–24.

DEFRA, (2006). *The macro-economic rebound effect of energy efficiency improvements and the UK economy*. Department of Environment, Food & Rural Affairs. http://www2.defra.gov.uk/research/project_data/More.asp?I=EE01015&M=KWS&V=Energy+Efficiency&SCOPE=0%20 [Accessed 11 October 2006].

Dimitropoulos, John (2006). *Energy Consumption Takes Time*. Proceedings of the 6th BIEE Academic Conference, Oxford UK, 20–21 September 2006.

EST (2006a). *Rise of the Machines: a review of energy using products in the home from the 1970s to today*. London: Energy Savings Trust. http://www.est.org.uk/uploads/documents/aboutest/Riseofthemachines.pdf [Accessed 17 October 2006].

Fouquet, Roger and Peter Pearson (2006). Seven Centuries of Energy Service: The price and use of light in the United Kingdom (1300–2000). *The Energy Journal*, 27/1: 139–76.

Ghazi, Polly and Judy Jones (2004). *Downshifting: The Bestselling Guide to Happier, Simpler Living*. London: Hodder & Stoughton.

Greene, D., Kahn, J. and Gibson, R., (1999). Fuel economy rebound effects for US household vehicles. *The Energy Journal* 20(3), 1–29.

Greening, A., Lorna, David Greene, C. Difiglio (2000). Energy efficiency and consumption – the rebound effect – a survey. *Energy Policy* 28(6–7), 389–401.

Heap, Brian and Jennifer Kent (eds) (2000). *Towards Sustainable Consumption: a European perspective*. London: The Royal Society.

Herring, Horace (2003). *Reducing Consumption Through Communal Living*. Paper to the ECEEE Summer Study 2003.

Herring, Horace (2005). Energy Efficiency: A Critical View? *Energy: the International Journal* 32(1): 10–20.

Hillman, Meyer and Tina Fawcett (2004). *How we can save the planet*. London: Penguin.

HoL (2006). *Energy efficiency*. First report of the session 2005–6. House of Lords Science and Technology Committee HL 21–1.

Hong, Sung, Tadj Oreszczyn, Ian Ridley and the Warm Front Study Group, 2006. The impact of energy efficient refurbishment on the space heating fuel consumption in English dwellings. *Energy and Buildings* 38/10: 1171–81.

Honoré, Carl, (2004). *In Praise of Slow: How a Worldwide Movement is Challenging the Cult of Speed*. London: Orion Press.

Hounsham, Stephen 2006. *Painting the Town Green: How to persuade people to be environmentally friendly*. London: Green-Engage/Transport 2000.
http://www.transport2000.org.uk/news/maintainNewsArticles.asp?NewsArticleID=286 [Accessed 17 October 2006].

Hub (2005). *Seeing the light: the impact of micro-generation on the way we use energy*. Report for the Sustainable Consumption Roundtable, 2005.

Jackson, Tim (ed.) (2006). *The Earthscan Reader in Sustainable Consumption*. London: Earthscan.

Ledbury, Matthew (2004). *UK car clubs: an effective way of cutting vehicle usage and emissions?* MSc dissertation, Environmental Change Institute, University of Oxford.

Levett, Roger, Ian Christie, Michael Jacobs and Riki Therivel (2003). *A Better Choice of Choice: Quality of life, consumption and economic growth.* London: Fabian Society.

McDonald, S., Oates, C., Young, W. and Hwang, K. (2006). Towards sustainable consumption: researching voluntary simplifiers. *Psychology and Marketing* 23/6: 515–34.

Rejeski, David, (ed.), (2003). E-commerce, the Internet and the Environment. Special issue of *Journal of Industrial Ecology* 6/2.

RCEP (2000) *Energy – The Changing Climate.* London: Stationery Office. http://www.rcep.org.uk/newenergy.htm [Accessed 17 October 2006].

RESOLVE (2006). The ESRC Research Group on Lifestyles, Values and Environment. University of Surrey. http://www.surrey.ac.uk/resolve/ [Accessed 17 October 2006].

SCR (2006). *I will if you will: Towards sustainable consumption.* Sustainable Consumption roundtable. http://www.sd-commission.org.uk/pages/020506.html [Accessed 17 October 2006].

Slavin, Terry (2006). Living in a dream. *The Guardian* 17 May 2006.

Smil, Vaclav (2003). *Energy at the Crossroads.* Boston: MIT Press.

UKERC (2007). [Rebound Project; Final report] http://www.ukerc.ac.uk/content/view/130/031 [Accessed 17 October 2006].

# Part II

# Making it Happen

# 8

# Supporting Renewables: Local Ownership, Wind Power and Sustainable Finance

*Dave Toke*

There are a range of ways to provide financial and institutional support for renewable energy projects – including direct government support and corporate investment. However, given that many renewable energy technologies are relatively small scale, there has also been enthusiasm for more grass roots orientated initiatives involving some degree of local involvement, including direct local ownership. In some countries, this approach has been adopted in the case of wind projects with some success. Quite apart from potentially contributing to local economic renewal, it is sometimes argued that local involvement, and especially ownership, helps to avoid local opposition to what may otherwise be seen as imposed projects.

Local ownership of wind power can mean several things. Often the definitions tend to be rooted in ideological theories, with co-operatives regarded as the purest and most desirable form. However, it is not the only useful form of local ownership. I use the term local and community ownership to mean the same things, although, as shall become clearer later, there are grounds to suggest that they do not always mean the same thing.

In this chapter I want to look at some of the theory and practice of locally owned wind power, some relevant theory, and practical ways it is implemented. I also want to investigate what sorts of conditions are most favourable to its development and what sorts of financial regimes are most suited to local ownership. Financial support for wind power is essential, no matter how the windfarms are owned. However it is also the case that differing financial conditions can affect local ownership in different ways. I discuss the financial issues in the context of the debate about the relative merits of support regimes that offer fixed electricity prices and long-term electricity supply contracts for wind power operators (Renewable Energy Feed-In Tariffs) and market-based systems such as the British Renewable Obligation.

I shall begin by explaining what is entailed by local ownership. I shall then discuss some theory which is relevant to helping us understand local ownership issues. Then I shall move on to look at some case studies in different countries. I shall then discuss some issues and controversies associated with local ownership of wind power before coming to a conclusion.

I shall now discuss the definition of local ownership. This will allow us to consider some of the opportunities and challenges surrounding local ownership. It should be said that these definitions are my own shortcuts which I feel helps introduce the subject rather than being a definitive technical guide to a topic that is subjected by a considerable degree of complexity by different taxation and incentive structures in different countries. I shall pick up on some of these issues when I describe country-specific cases.

## Local co-operatives

Local wind power co-operatives have gone down in wind power folk lore as the political 'holy grail' for the green and left wing idealists interested in renewable energy. A local wind power co-operative, most famously implemented in Denmark in the 1980s and 1990s, means that the wind turbine or windfarm is owned, lock, stock and transformer, by local residents. Locally owned wind power co-operatives are also very common in Germany. They are called 'burgerwindparks'. The method of investing in wind power co-operatives will vary. In Denmark the co-operatives have been financed entirely out of equity capital, with no money borrowed by the bank. On the other hand, in Germany, the 'burgerwindparks' follow the standard project finance pattern of the bank lending the large bulk of the capital, with the shareholders, or equity portion, providing around 20% of the capital for the project.

## Non-local and non-commercial co-operatives

Even though this chapter is about local ownership it is necessary to point out the distinction between local and non-local co-operatives. Many windfarms are owned by collections of private investors who are not local, or at least not necessarily so. Some of these, for example Dutch wind power co-operatives are best described as ethical investor operations. Other examples, sometimes in the US and particularly in Germany, exist as means of relatively high income earners a good profit from their investments in windfarms. As we shall see in Germany the commercial co-operatives can also be described as 'corporate' ownership, a practice I have followed in Figure 8.1 later in the chapter.

## Farmer ownership

The majority of onshore wind power capacity in Denmark is farmer-owned, and it is a common practice in other places. Usually farmers will own a limited liability company, perhaps in collaboration with other farmers, which will own the windfarm. Farmers will usually borrow the bulk of the money needed to capitalise the project from a bank. It is occasionally the case that a non-farmer will own a wind turbine on their land, but farmers are the most common form of landowner. In theory many farmers are in a good position to invest in wind power since they have assets against which they can borrow and the assurance of a site for the windfarm. However, motivation to require the necessary knowledge is often the key barrier to farmer ownership of wind power.

## Local ownership and planning acceptability – some theory

So how, in theoretical terms, might local ownership improve the possibilities of local acceptance of wind power schemes? One suggestion is that a farmer or local co-operative members can utilise his/her own social network to advance the planning case.

Personal links to local actors can make a subtle but significant difference to the prospects of gaining planning permission. Most of all, a local farmer is in a much better position (because they have the social contacts) to organise a campaign in favour of a windfarm proposal compared to an outside developer who does not have very good local contacts.

This type of explanation can be linked to a social capital explanation of wind power planning outcomes and the ways in which locally inspired schemes increase the chances of planning acceptance. Social capital can, among other ways, be calculated in terms of the number and strength of contacts and networks enjoyed by an individual (Putnam, 2000). The more of this that is possessed by an actor, the more ability they have to influence local political decisions, especially in windfarm planning cases where the opinion of people in the immediate vicinity is crucial to the outcome (Toke, 2005a). Windfarm schemes that are actively promoted by local owners are, according to this theory, much more likely to be approved. One study does suggest that wind power planning outcomes are influenced by the relative strength of pro-windfarm or anti-windfarm planning networks (Loring, 2004).

Put in simple terms local ownership may counteract what is often called the 'NIMBY' tendency, although convincing academic work has been done to contest the notion of classifying opposition to windfarms as 'NIMBY' (Wolsink, 2001). In practice the large majority of people who oppose windfarms say they oppose all windfarms, not just ones near where they live. Even this rejection of NIMBYism may itself be too simplistic and based on a 'positivist' view of data. The answer to questions about attitudes to windfarms in general has to be understood in the context of a specific proposal. In addition the respondents are likely to want to avoid the 'NIMBY' accusation and to 'universalise' their objections to the windfarm proposal (Haggett and Toke, 2006).

The point is that people's perception of windfarms is likely to be significantly influenced by how they see their own identity, which is place-specific. Moreover, those living close to a proposed windfarm may see that taking action to oppose the windfarm proposal may be worthwhile to eliminate what they see as a significant threat to their visual amenity (Toke, 2002).

There is, in fact a theoretical literature on place identity (Wester-Herber, 2004, Twigger-Ross and Uzzell, 1996). Place identity is said to consist of four aspects: distinctiveness which means that the place will evoke particular connections between the individual and what they see as being their preferred identity; continuity which can connect an individual to the past; self esteem which can be concerned with actively maintaining the place itself; and self-efficacy which can refer to living in a 'manageable environment' (Twigger-Ross and Uzzell, 1996: 208).

Some will see windfarms as a threat to their sense of place identity, especially as they involve change. However, it is also possible to theorise on how local ownership may enhance, or at least ameliorate threats to, place identity. Twigger-Ross and Uzzell (1996: 207) report talks of evidence from psychology literature on how 'Evidence that having control, or not, over the maintenance of continuity of place is important to psychological well being'. In addition, the derivation of income from windfarms owned by local people can bolster their notion of 'self-efficacy'.

## Patterns of wind power ownership

Figure 8.1 indicates that the majority of wind power capacity in Germany is owned by 'corporate' interests. This is somewhat misleading in the sense that the corporate interests involved are rather different compared to the windfarms in Spain and the UK which are mostly owned by offshoots of major multination electricity companies.

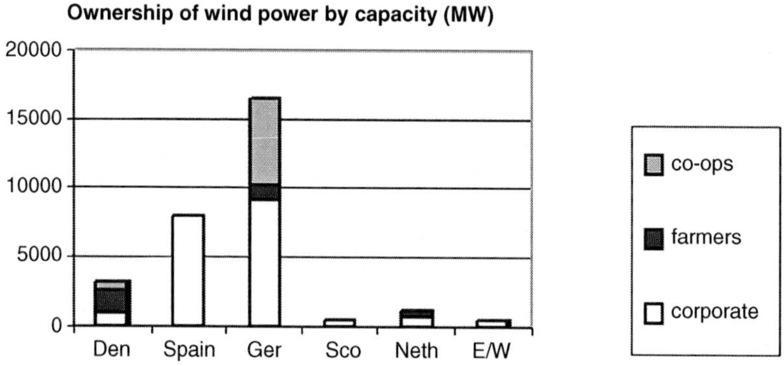

**Ownership of wind power by capacity (MW)**

20000
15000
10000
5000
0
Den Spain Ger Sco Neth E/W
co-ops
farmers
corporate

*Figure 8.1* Patterns of ownership of wind power in selected European states (end of 2004)

*Source*: Toke, D., Wolsink, M. and Breukers, S., 'Wind power deployment outcomes: How can we account for the differences? '*Renewable and Sustainable Energy Reviews*, forthcoming.

*Note*: the Danish statistics include offshore wind installations constituting around 14% of installed capacity, most of which is owned by (corporate) utilities. Danish onshore wind power owned by the corporate sector constitutes 12% of all wind power capacity.

*Original sources of data*: Interviews with: a) Alfonso Cano, APPA (Association of Producers of Renewable Energy), (Spain) 9/12/2004 b) Henning Holst, Independent Wind Power Consultant (Germany) 18/7/1999, and giving data from Enercon in Presentation to Conference, 'Locating Renewables in a Community Context', Open University, Milton Keynes, 15/11/05; c) Per Nielsen, Director, Energi Mileu Data (Denmark) 10/08/2004; d) Susanne Agterbosch, Utrecht University (Netherlands) 10/02/2005; e) figures from: www.bwea.com

As can be seen in Figure 8.1, the largest amount of locally owned wind power is in Germany. Please read the section of Germany to see the distinctions between different forms of ownership, but in this Figure 'co-operatives' in the German case means local windfarm co-operatives. The totals in Figure 8.1 are now dated. Even by the summer of 2006 there was more than 18,000 MW of wind power in Germany and over 10,000 MW of wind power in Spain.

I do not say much about Spain in this Chapter, but Spain is an interesting case insofar as there are no real examples of local ownership in Spain. However, the circumstances in Spain are different from the other cases for several reasons. First, rural Spain is subject to poverty and continuing depopulation. The countryside is rarely seen as a valuable resource to be defended from visual intrusions. Moreover, local municipalities can look forward to earning considerable amounts in taxation from wind power. There are no organised groups campaigning to preserve landscape, so the local environment is (more often than not) very conducive to wind power development – arguably much more so than in the other cases in the Figure 8.1. It can therefore be argued that local ownership would not add much to the supportive environment for wind power anyway.

The main utilities in Spain are keen to develop wind power in the context of a rapidly increasing demand for electricity and few indigenous energy sources. At the end of 2005 wind power supplied 10% of Spain's needs, and this proportion is rapidly increasing.

However, in many, if not the majority, of the areas in developed countries, there are strong arguments that local ownership will bring significant benefits, both to the aim of increasing the amount of wind power installed and also in increasing support for the technology. Denmark is regarded as the birthplace of the modern wind power movement.

## Denmark

Wind power largely developed as a modern technology in the late 1970s and early 1980s as a result of grass roots technological innovation in the Danish agricultural sector (Karnoe, 1990). Danish wind power now supplies almost 20% of Danish electricity consumption. This comes from around 3,500 MW of wind power capacity, around 3,000 MW being locally owned.

Most local ownership is bound up with small projects. This has been source of confusion in the UK, as some have assumed it is the small size rather than the local ownership that has given locally owned schemes greater planning acceptance compared to larger, utility-owned projects.

There were regulations aimed to prevent profiteering by remote investors, and so keep (non-utility) ownership local. The principle of local ownership was enshrined in Danish Planning Law in the 1980s. The law prescribed three forms of ownership; by utilities, by co-operatives owned by local people or by farmers (who were limited to one machine each). People could only own wind turbines if it was placed in their own Kommune or the one next to it. These had the effect of keeping non-utility schemes small. These restrictions were ended in 2001, but only at the same time as the subsidies for new onshore wind power schemes were terminated.

The co-operative institution was instrumental in launching the Danish wind power industry, and with it the wind power industry in the world generally. Few argue with this proposition, although many now argue that the wind industry should abandon its romantic past and focus on becoming part of the mainstream, multinational, electro-engineering industry. The grass roots basis of the development of the Danish wind power sector was very successful in that something like 5% of all Danes became shareholders.

The development of Danish community wind power was facilitated, in terms of incentives, because there were tax concessions given to wind power which allowed windfarms to be built with the assurance that they would receive a secure income for an extended period. This was done by what is known as a 'Renewable Energy Feed-In tariff' (REFIT). The REFIT was terminated at the end of 2001. The only feed-in tariff is now for offshore wind power and also for the re-powering of existing onshore schemes. The curtailment of the REFIT was in the context of two factors. First, there was public pressure to contain increases in electricity prices that were a consequence of the incentives for wind power. The second factor was the election of a right wing Government in November 2001.

The main incentive for the Danish programme was that owners of shares in windfarms in effect earned the tax that would normally be levied on the electricity generated by the windmills. All electricity is subject to significant taxes in Denmark. This allowed wind power operators to be paid more or less the same level as the domestic price of electricity. There was a system developed with local banks whereby money borrowed by co-operative members was repaid using tax rebate receipts that were, by arrangement, handled by the banks.

The electricity utilities caught onto the coat-tails of the co-operative wind power movement and sold shares in their schemes as well. As the 1990s wore on more and more farmers invested in wind turbines. Such was the commercial pressure that the farmers kept the best sites to themselves, or sold the sites for high prices, so that the number of co-operatives being established dwindled in comparison to farmer-owned wind power.

It would, however, be wrong to try to explain the proliferation of community wind power in Denmark by reference to the REFIT system alone, whatever the merits of that system may be. If one wants to explain the success of the Danish wind power movement it is necessary to study the social movement that propelled wind power forward as a technology in order to properly understand the differences between the development of Danish wind co-operatives and what has happened elsewhere (Jameson *et al.* 1990, pp. 66–120). The renewable energy movement in Denmark was a grass roots mass political movement springing from a mass, active anti-nuclear movement and a Renewable Energy Organisation which itself involved over 0.5% of the Danish population. That would be about 300,000 of the British population or around 1.5 million in the US. Both countries could certainly do with that many grass roots activists. Then we would have wind power co-operatives! A financial system that suited them would follow.

## United Kingdom

Community wind power has been slow to develop in the UK, despite the existence of some financial incentives and a planning regime favourable to wind power development since 1990. When community wind power did make some tentative roots, in the form of the Baywind Co-operative, it was as a result of an initiative from Sweden. After establishing a scheme in Cumbria the Baywind Co-operative formed 'Energy 4 All, a company concerned with working with developers to sell windfarm shares on an ethical investment basis to the general public.

There are some co-operatives being planned in Scotland, as well as a small one in Wales, and there is a string of planned windfarms which are to be owned by ethical investors (organised by Energy4All) who are not necessarily local. Ethical investment share offers have so far been fully subscribed. In 2006 one raised over £3 million for the Westmill project in Southern England. This is the brainchild of a farmer-landowner called Adam Twine. Around £700,000 has been raised for the Boyndie windfarm near Banff. In this latter case most of the community investment is coming from Aberdeenshire itself, and this will mean that a windfarm that is otherwise financed by conventional corporate means will be part-owned by community investment. Energy4All has organised this strategy with Falck Renewables. Energy4All is also co-operating with Wind Prospect who are developing various projects in the East of England that will be part-funded by community investment through 'Fenland Green Investments' who are organising the share offers.

There are also a small number of examples of windfarms owned by farmers. Two schemes being currently developed in Wales, the Moel Maelogen and Amegni schemes, are both owned by groups of farmers and both comprise 15.6 MW projects. However, all of this activity will not amount to more than 50–60 MW of wind power owned by community means. This compares to around 3,000 MW of wind power that is likely to be in place in the UK by the end of 2007.

There have been discussions concerning reform of the Renewable Obligation to give a specific 'band' in the obligation target specifically to community wind power. As I write, the RO is in the process of being reformed, although the principal concern is to make it more cost-effective and provide more income for offshore windfarms rather than onshore schemes. However the problem is that there are so few people actively engaged in promoting community wind power initiatives in the UK. If there is only a small lobby then politicians may not produce appropriate policy responses. Unlike Denmark or Germany there has never been a major grass roots energy movement in the UK. Some people may comment that we do have an anti-nuclear movement in the UK, but they often do not realise that this pales in comparison to the activism and protest associated with anti-nuclear movements in Denmark and Germany.

There have been some 'alternative energy' networks in the UK, perhaps the best known being the Centre for Alternative Energy (CAT) In Wales. However, it may also be significant that when the first intimations of a British Renewable Programme emerged at the beginning of the 1990s some key CAT activists either decided to look for multinational financial backing for large, remotely owned, wind power projects or formed a wind power consultancy and services company. This is as opposed to trying to proselytise about community wind power.

There are a small number of organisations, such as Energy4All and Baywind Co-Operative attempting to stimulate activity. There is also a limited number of actual and wannabe farmer-owners, but outside of this there is relatively little determination to overcome obstacles that face all wind power developers. I must say that in contrast to other countries covered in this study an atmosphere of 'can't do' is pervasive in relation to starting locally owned wind power schemes. The real and imaginary failings of the British RO may be wheeled out as an excuse for this. However, in spite of many criticisms of the British RO and the somewhat simplistic talk of its alleged 'riskiness' for windfarm developers, there are actually secure, long-term electricity supply contracts available for all types of wind power operators paying £50–£60 per MWh (Toke, 2005b). This is a good price for British wind conditions, and these terms are better than exists in most countries.

The RO can certainly be, and ought to be, improved to favour community wind power. However, the biggest obstacle is the attitudes and culture that prevails in the UK, not the technicalities of the British 'market-based' financial support regime.

## Germany

There will be approaching 10,000 MW of locally owned wind power in Germany out of nearly 20,000 MW that will be in place by the end of 2006 (extending data from Figure 8.1). This will produce around 6% of German electricity consumption. In contrast to Denmark, many of the locally owned schemes are quite large. Schemes have tended to be built up incrementally, up to 30–50 MW in size.

The expansion of German wind power has been supported by a REFIT scheme that has been operating at a Federal level since 1990. Premium tariffs for electricity from wind power and other renewable technologies are guaranteed for each scheme for a 20 year period. However, there have been increasing political pressures on this REFIT, a lot of them orchestrated by the monopoly utilities who have constantly complained of the extra costs needed to pay for windfarms. Electricity demand has been stationery since the country was reunified, and the utilities do not seem to like competition with their existing coal and nuclear power stations.

It should be borne in mind that the level of the feed-in tariff in Germany has, as a result of the 2004 Electricity Law, declined to such an extent that such schemes would not be at all profitable in similar wind conditions in the UK. *Wind Power Monthly* has suggested that co-operative wind power is going out of fashion in Germany. The truth is different (Knight, 2004). What is not mentioned very much by *Wind Power Monthly* is that the tail-off in German wind power development has been much slower than some predictions precisely because the idealism of the ethical investors in 'burger windparks' has allowed development to continue at lower rates of return than the corporate windparks.

Then, commenting on the lack of new wind power in Denmark, in March 2005, Torgny Moller said (also in *Wind Power Monthly)* that 'the tradition of community owned wind plant that dominated the market ten years ago has all but disappeared' (Moller, 2005). By apparently emphasising a decline in local and co-operative ownership of wind power *Wind Power Monthly* gives the impression that corporate developers have lower costs and better cheaper means of investing in wind power. The fact that there is little current onshore wind power deployment in Denmark is in the context of faltering levels of subsidies, not because of a surfeit of amateurism at a local level. The political benefits for local ownership have been clearly established. As mentioned previously, the corporate bodies that have developed around half of wind power in Germany are themselves co-operatives in the sense that the equity capital is raised through public share offers rather than through institutional investment that is the rule in the UK and Spain.

The corporate wind development companies are known as *kommandistengedellschaften*. These companies derive their capital from high income earners. In Germany the tax rules allow people to offset their taxes against investments, making such investment potentially profitable for people on high marginal tax brackets. However, the German government has altered the tax rules meaning that investors have to wait longer to receive the same income on their capital compared to the previous rules. Because the high income investors in wind power are principally interested in financial returns, this part of the wind development market has faltered. Declining subsidy levels and changes in tax systems have stilted their operations. However, a switch to use of bonds to finance windfarms is being used by the corporate sector to circumvent this problem (Weinhold, 2006, p. 43).

Germany is another country (besides Denmark) that boasted (still boasts) a militant anti-nuclear movement. Like Denmark, and unlike the UK, there have been many mass anti-nuclear protests, and in Germany a lot of them have involved violent confrontations with the police over nuclear waste dumps (again, unlike the UK). There have also been many renewable energy activists in Germany who have formed co-operatives. In fact, these days, the co-operatives in Germany have a larger proportion of the wind power generation market than is the case in Denmark where the majority of onshore wind power is owned by farmers. This can be seen in Figure 8.1. These local co-operatives are so-called 'burgerwindparks' which consist of ethical investors who want wind power to be developed locally.

The ethical wind power co-operative movement's strength is evidenced not only by a continuing desire to invest in onshore wind power, but also attempts to invest in offshore wind power. There is already considerable co-operative investment in Danish offshore wind power, but a very large project co-operative-financed project is being planned in Germany. A number of organisers of *burgerwindparks* are collaborating to develop a 240 MW offshore windpark off the north coast of Germany near the Island of Sylt. Interest has already been shown in this, *Budentiek,* project from over 8,000 potential investors. However, the project has, for the moment, like many other offshore projects around Europe, by the jump in prices of wind turbines. The accelerating demand for wind power has not been matched by sufficient wind generator manufacturing capacity, partly because of the 'boom and bust' nature of the US wind power market. We shall move onto this topic now.

## North America

Given the US's reputation as a country dominated by corporate interests, you would not expect any interest in co-operative or even local ownership of wind power. However, this view is simplistic, at least in the case of some states. Minnesota is the leader in terms of local ownership of wind power. By the middle of 2006 there was around 150 MW of locally owned wind power installed, around half of that being farmer owned and the rest being a mixture of individuals, schools and municipally owned schemes. Farmers tend to establish schemes which are owned by what in effect amounts to farmer co-operatives.

In the US financial support for wind power comes through a mixture of a Federal Production Tax Credit (PTC) and state-based 'renewable portfolio standards' (RPS). These Renewable Portfolio Standards exist in around 20 states and they mandate electricity suppliers to provide a stated proportion of their electricity through renewable energy sources by a stated date. However, little wind power development takes place either outside states where there is an RPS or, for that matter, during periods when the PTC has not been available.

The nature of the Federal Production Tax Credit (PTC) incentive (now worth 1.9 cents/Kwh) means that companies or individuals with large tax liabilities have to be the main investors. The production tax credit creates problems for the global wind industry as well as the US industry in the sense that it is intermittently turned on or off by Congress. This produces 'boom or bust' wind power development conditions which has unfortunate consequences for the world wind power market. In the 'boom' period when developers are rushing to install windfarms there are not enough wind turbines being manufactured, meaning a big increase in wind turbine prices. The most expensive schemes, especially offshore windfarms, are thus delayed. Because there is uncertainty about whether the US wind market will continue after the current PTC is ended in December 2006, wind generator manufacturers are reluctant to extend their production lines.

The nature of the production tax incentive can create specific difficulties for locally owned schemes, although some doubt that this will prevent farmers from investing in wind power. According to Mark Willers, the prime mover behind a 3.8 MW project in Buffalo Ridge, Minnesota:

> [T]he most difficult step in these projects was not finding capital for the hardware, consultants and legal fees because farmers were enthusiastic about investing from the very beginning. He believes that it is a myth that farmers do not have the money to finance projects on this scale (Minwind I and II will cost about $1.6 million dollars each and will be paid off in ten years). The biggest obstacle, rather, was negotiating a power purchase agreement, a crucial step to moving any wind project forward (Windustry, 2002).

Certainly, feed-in arrangements will guarantee the offer of a power purchase agreement. In order to achieve, and make use of such arrangements there needs to be a number of people interested in setting up community wind power schemes in the first place. However there is a co-operative tradition in Minnesota, added to which there is recent experience of forming co-operatives to produce ethanol from corn. Farmers can pool their investments to mobilise the production tax incentive and provide equity finance for the wind projects. This is what has led to farmer-ownership of wind power development in Minnesota.

The state of Minnesota has a programme of expanding community wind power. The State Governor, Tim Pawenty, has established a target of installing 1,000 MW of community wind power by the year 2010. The community wind power scheme offers around 3 cents/Kwh to locally owned schemes in addition to the Production Tax Credit incentive. On top of community wind power Minnesota already has 640 MW of existing wind power owned by corporate interests.

Minnesota has the largest community wind power programme in the US. There is some state governmental support for wind power in Iowa, Montana and Oregon. However, despite this progress no more than around 2% of the 10,000 MW of wind power installed in the US by summer 2006 is community owned.

There are also strong co-operative traditions in parts of Canada, including Ontario. The Ontario Government announced in 2006 that there would be 'standard offer' contracts which would be broadly similar to a feed-in tariff, operating for projects up to 10 MW in size. The Ontario Sustainable Energy Association (OSEA) has been campaigning for this policy (http://www.ontario-sea.org/). Community wind power initiatives are being organised, in particular by the Toronto Renewable Energy Co-operative (TREC). TREC has a target of establishing around 100 MW of locally owned wind power in the next few years. There is a great deal of enthusiasm for wind power in Canada, but the wind industry lacks the sort of planning regimes and financial support systems that are, or have been, in evidence in European countries that I discuss in this chapter and elsewhere (Toke, 2005c).

## The Netherlands

Enthusiasm for wind power in the Netherlands emerged at the end of the 1980s in the form of a co-operative wind power movement. A dozen wind power co-operatives have put together projects which now constitute 5% of total wind power capacity in the Netherlands. These co-operatives consist of ethical investors but they do not necessarily come from the same area as the windfarm itself. In recent times the wind power co-operatives have tried to develop in co-operation with mainstream wind power developers. More details can be derived from the Dutch wind co-operatives website http://www.duurzameenergie.org/odewindcoop.html. In total, wind power generated around 3% of Dutch electricity in 2005.

There is also a big farmer-owned windfarm sector. Indeed 40% of Dutch wind power is owned by farmers, mainly in Flevoland and Noord Holland. These are areas which have been reclaimed from the sea. Certainly farmers living on reclaimed land in the Netherlands and also around the coast near Husum in North Germany seem to be especially enthusiastic about using the land for wind power. Perhaps this fits in with an attitude that sees reclaimed land as a new resource which needs to be utilised. The wind power industry in the Netherlands has had to cope with a changing incentive system. Currently there is a type of REFIT in place. This superseded a market-based 'green electricity certificate' scheme which was wound up because of design flaws.

## Discussion

I want to set this discussion in context. It does seem to me that we have two ideological polls in the wind industry. One, perhaps represented by authors such as Frede Hvelplund (2005), is critical of the role of the major energy utilities, and sees local ownership as the most important strategy for a ecologically sustainable wind industry. Another, perhaps typified by Marcus Rand, the former Chief Executive of the British Wind Energy Association (BWEA), sees integration into the mainstream energy industry rather than reliance on association with environmental ideals as the preferred future strategy for wind power (Rand, 2005). Many would see the BWEA as making common cause with the American Wind Energy Association as being more sympathetic to a corporate-led wind power industry rather than a German-Danish vision of an independent, popularly owned, industry.

I would disagree with those who seek to emphasise a polarity between these two supposedly different camps, although I do very much respect the people who hold these views as people doing what they see as their best to support the technology. It is important to say that limiting ones vision either to solely locally owned or solely corporately owned windfarms is to limit wind power as a technology. We want more wind power, not less, and ruling out one or other type of ownership of wind power means fewer installed gigawatts and a lot less clean energy.

It is not a necessary condition for a widespread community or local ownership to have undying hostility between a grass roots wind power movement and the corporate interests who dominate the electricity industry. On the other hand it does seem to me to be patently foolish to go around saying that the wind power industry should no longer emphasise an idealistic 'environmental' image. It is also short-sighted to insist that the whole wind power enterprise must be bought up and run solely by large multinational companies.

Subsidies for renewable energy will dry up without public support, and renewable energy supporters must not believe their own propaganda to the extent that they believe renewables will soon not have need of subsidies. It will not matter how many friends you have in the boardrooms of Siemens, E.On, RWE or Centrica if there is not a big enough income stream to sustain wind power and other renewable energy technologies. The electricity majors will not invest in loss-makers. It is necessary to maintain public support for subsidies for wind power. This will be aided by appeals to both environmental idealism and developing community involvement in, and especially ownership of, wind power schemes.

Many look forward to the day when subsidies will no longer be needed for wind power. I do not doubt that this can happen some day, but there is still a long way to go before capital intensive technologies can easily compete on short-term prices with gas-fired power plant. Current high oil and gas prices may give an illusion of competitiveness for wind power, but the experience of the slump in energy prices after the 1970s oil crises should make us wary of assuming that the current high fossil prices will remain at their present level. These levels may merely be a result of the current state of the energy investment cycle which will bring prices crashing down a few years hence.

It makes obvious sense to encourage as much grass roots support and involvement in the wind industry as is practically possible, especially in cases such as the UK where there is a well organised and very active opposition to onshore wind power. It is stupid and short-sighted to dismiss the promotion of local ownership. The local opposition to windfarms that is so evident in the UK, and which puts itself forward as the 'real' voice of ordinary people needs a strong grass-roots response if we wish to avoid a more rapid deployment of wind power. This is additional to the persuasive argument that there is actually a lot to be said, in ecological-political terms, in support of the principle of local ownership of resources in general (Hines, 2000). The major energy companies should be encouraged to support and engage in renewable development, but it is the voters, not the multinationals, that is the bulwark upon which subsidies for renewable energy rest and continue to be made available.

On the other hand it is very short-sighted not to support a windfarm merely because it is owned by corporate interests. Will such a strategy cut carbon dioxide emissions? Will opposition to corporate efforts to develop wind power reduce arguments for new nuclear power stations? Of course it will not, for the practical effect of a failure to support corporately financed renewable energy will be to reduce the amount of renewable energy, not increase it.

I must confess that a few years ago I was one of the people who argued for a 'community wind power only' strategy in the UK (Toke and Elliott, 2000). This was, it has to be said, at a period when there was practically no wind development activity at all and the wind power industry seemed to have made an absolute mess of the opportunities (such as they were) presented by the then renewable energy support system, the Renewable NFFO. However, more recently, I can see that there is quite a lot of development going on under the Renewables Obligation. While I would have preferred to start off with a 'Renewable Energy Feed-In tariff' (REFIT), I do not accept that there is something inherent in the RO that prevents locally owned schemes happening.

I have discussed, earlier, how much continental European wind power is locally owned. If not for local ownership we would have much less wind power on the European continent. It is also the case that if there is a strong lobby for local ownership then we are likely to see systems that are more convenient for local owners. Hence it is much more likely to be a case of 'local ownership lobby leads to REFIT' rather than 'REFIT leads to local ownership'. We have a REFIT in Spain, but there is even less local ownership than in the UK or the US.

## Conclusion

Local ownership takes various shapes and sizes, and, as we can see in the case of public wind investment co-operatives, not all of the non-corporate types are actually locally owned. Local ownership builds on local social capital to achieve greater acceptance for wind power schemes, and it can also increase support by returning profits from the wind power to the local communities. In so-doing, the identification of inhabitants with their locality can be enhanced rather than reduced. However, even non-local public investment can be important since it gives a wide number of people a stake in wind power, and therefore bolsters the political lobby in favour of the technology.

It is possible for locally owned schemes to flourish under different types of financial support system, although it important to ensure that non-corporate developers are offered financial terms and electricity supply contracts that are at least as favourable to the corporate projects. A REFIT system seems to be the best means of doing this. However a lot of this may have something to do with the probability that REFIT systems are best for all types of developers, not just the locally owned or cooperative schemes!

My feeling, however, is that it is wrong to say that the existence of a REFIT system will ensure local ownership, and equally wrong to say that 'market-based' systems necessarily allow only corporate concerns. Rather, a different analysis is more useful. Strong political lobbies from farmers and co-operative wind enthusiasts tend to generate terms and conditions that favour locally owned schemes. It is the cultural aspect of support for local ownership, based often on an existing co-operative or local energy activist tradition, that is most important for developing local ownership. Wind enthusiasts would do well to concentrate more effort on developing this culture of local energy activism rather than bicker amongst themselves. Certainly the Canadian Wind Energy Association is an example of a trade association that appears to be looking towards a broad church approach that wants to maximise both corporate and community wind energy strategies.

## Web references

**Germany**: Budentiek co-operative offshore windfarm (2005) http://www.butendiek.de/

**England and Scotland**: 1. Energy4All: http://www.energy4all.co.uk/ 2. Baywind Energy Co-operative: www.baywind.co.uk 3. Community Wind Power Network http://www.freewebs.com/communitywindpowernetwork/

**Wales**:

1. Awel Aman Tawe proposed community windfarm: http://www.awelamantawe.org.uk/
2. Moel Maelogen windfarm: http://www.ailwynt.co.uk/
3. Dyfi Eco Valley Partnership: www.ecodyfi.org.uk

**North America**:

1. Minnesota: Windustry; http://www.windustry.com/
2. Ontario: Ontario Sustainable Energy Association http://www.ontario-sea.org/

**The Netherlands**: http://www.duurzameenergie.org/odewindcoop.html

## Bibliographic references

Haggett, C. and Toke, D. (2006) Crossing the great divide – Using multi-method analysis to understand opposition to windfarms', *Public Administration*, Vol. 84, No. 1, 2006 pp. 103–20.

Hines, C. (2000) *Localization: A Global Manifesto*, London: Earthscan.

Hishelwood, E. (2000) 'Community Funded Wind Power – The Missing Link in UK Wind Farm Development?', *Wind Engineering*, Vol. 24, No. 4, pp. 299–305.

Jameson A., Eyerman, R., Cramer, C., Læssoe, J. (1990) *The Making of the New Environmental Consciousness. A Comparative Study of the Environmental Movements in Sweden, Denmark and the Netherlands*, Edinburgh: Edinburgh University Press.

Karnoe, P. (1990) 'Technological Innovation and Industrial Organisation in the Danish Wind Industry', *Entrepreneurship and Regional Development*, Vol. 2, No. 2, pp. 105–23.

Knight, S. (2004) 'German Financing Revolution', *Windpower Monthly*, Vol. 20, No. 11, (November), pp. 59–62.

Loring, A. (2004) *Wind development in England, Wales and Denmark – The Role of Community Participation and Network Stability in Project Acceptance and Planning Success*, D. Phil thesis, Falmer, Sussex: University of Sussex.

Moller, T. (2005) 'Worst year for wind in two decades', *Windpower Monthly*, Vol. 21, No. 3, (March), 48.

Putnam, R. (2000) *Bowling Alone*, New York: Simon and Schuster.

Rand, M. (2005) Speech to General Meeting of British Wind Energy Association, December 7th, BWEA offices, London.

Toke, D. and Elliott, D. (2000) 'A Fresh Start for Wind Power?', *International Journal of Ambient Energy*, Vol. 21, No. 2, pp. 67–76.

Toke, D. (2002) 'Wind Power in UK and Denmark: Can Rational Choice Help Explain Different Outcomes?', *Environmental Politics*, Vol. 11, No. 4, 2002, pp. 83–100.

Toke, D. (2005a) 'Explaining wind power planning outcomes, some findings from a study in England and Wales', *Energy Policy*, Vol. 33, Issue 12 (August 2005), pp. 1527–39.

Toke (2005b) *'Are green electricity certificates the way forward for renewable energy? An evaluation of the UK's Renewables Obligation in the context of international comparisons', Environment and Planning C*, 2005 Vol. 23, No. 3, pp. 361–74.

Toke (2005c) 'Community Wind Power in Europe and in the UK', *Wind Engineering*, Vol. 29 No. 3 pp. 301–8.

Twigger-Ross, C. and Uzzell, D. (1996) 'Place and Identity Processes', *Journal of Environmental Psychology*, Vol. 16, 205–20.

Weinhold, N. (2006) 'Vestas Powering Germany', *New Energy*, August 2006, pp. 42–5.

Wester-Herber, M. (2004) 'Underlying concerns in land-use conflicts-the role of place identity in risk identity in risk perception', *Environmental Science and Policy*, Vol. 7, 109–16.

Windustry (2002) 'Minwind I and II: Innovative farmer-owned wind projects' Fall 2002 Newsletter https://www.windustry.com/newsletter/2002FallNews.htm, accessed August 2006.

Wolsink, M. (2000) 'Wind Power and the nimby-Myth. Institutional Capacity and the Limited Significance of Public Support', *Renewable Energy*, 21 (1) 49–64.

# 9
# Supporting Renewables: Feed in Tariffs and Quota/Trading Systems

*David Elliott*

New technologies, like renewable energy systems, usually need support to become established. Some governments have focused on initial direct state aid in the form of grants (e.g. for R&D) and then direct operating subsidies for new projects, while others have sought to make use of market mechanisms, including emission trading and green certificate systems, in order to bring emergent technologies to the market while keeping prices down. The latter approach is represented by the EU Emission Trading System and the UK's Renewable Obligation (RO). However, most EU countries have adopted variants of another approach, Renewable Energy Feed in Tariffs ('REFIT'), first used in Germany, which provides a guaranteed market with fixed prices. This has been clearly more successful so far than the competitive market/trading approach in terms of delivering much more generating capacity – for example, despite having a much worse wind regime, by 2004 Germany had installed over 20 times more wind capacity than the UK. And, perhaps surprisingly, it has also cost less. This chapter explains why.

## Supporting renewables

In 2001, the UK Government initiated a wide ranging review of energy policy which looked decades ahead and concluded that, of the main options likely to be available for responding to climate change, renewable energy and energy efficiency looked to be the best bets. Nuclear power was relegated to an insurance role – to be called on if the favoured options did not deliver as expected (PIU 2002).

This was a comprehensive Cabinet Office-led review, which drew on 'learning curve' analysis to assess the economic prospects of the various energy options. Learning curves plot performance (usually in terms of cost/kWh) against volume of production (reflecting the number of systems in use, usually using total kWh delivered or kW capacity installed). With most technologies, as experience is gained with new technologies and as markets expand for them, economies of scale and market pressures kick in, and costs fall. In terms of renewables, it was found that, whereas wind and solar PV technologies had demonstrated quite rapid progress down their learning curves, with the slope of the learning curves for 1980–95 (measured in log kWh costs/log kW volume) of 18–20%, nuclear technology was still only making very slow progress, with a learning curve slope for 1975–93 of just 5.8%.

Given that, for New Labour, economics was deemed to be all-important, the choice was clear. And that is what emerged in the 2003 Energy White paper – a commitment to renewables along with energy efficiency, with the nuclear option left open, but in effect sidelined on strategic/economic grounds.

## Renewables botched

All could well have been fine if it was not for the continued preoccupation with using competitive market mechanisms to achieve the price reductions predicted by the learning curves. Labour did this by imposing a Renewables Obligation (RO) on energy supply companies – they had to get fixed percentages of their electricity from renewable energy sources, expanding in stages to 10% by 2010, later expanded to 15% by 2015. Given the emphasis on specified capacity targets, the RO is therefore sometimes called a 'quota' system. The level of the RO, and the level of 'buy out' price that was introduced for companies that did not meet the obligation, was chosen in part to ensure that the extra cost of meeting it, which the suppliers could pass on to consumers, remained relatively low. The expectation was that the overall electricity price increase between 1990–2010 would be about 5.7%. But it was up to the companies to decide how to meet the targets/quotas within a competitive market framework, and naturally they choose the cheapest options – initially sewage gas and land fill gas, but increasingly wind power.

However, as a the result of the highly competitive market created by the RO arrangements, only a few new projects have gone ahead – and the 10% by 2010 target began to look unlikely to be achieved. By comparison, Germany and many other EU countries had adopted a different, arguably much simpler, direct subsidy approach, which has proven to be far more successful than the competitive price/quota approach. For example, by 2004, the UK had only managed to install about 600 MW of wind generation capacity while Germany had installed over 12,000 MW – more than 20 times more. And this in a country with a far worse wind regime – much lower average wind speeds – than the UK.

How was this done? Germany's system, initially called the 'Renewable Energy Feed-in Tariff' (REFIT), provided guaranteed fixed-price subsidies for renewables, and that created a secure investment climate for expansion. By contrast, under the RO, in the UK prices were determined by the market. To provide an incentive for compliance, suppliers were given Renewable Obligation Certificates (ROC's) for each eligible MWh of renewable electricity, and they could trade any excess ROC's they accumulated, over and above their allocated RO quota, with suppliers who had not managed to reach their quota. This trading aspect is why the RO is sometimes referred to as a 'quota/trading' system. The value of the ROC's would of course be variable, being determined by the market for them created by the RO system, and the companies success in operating within it.

It was argued that, although the REFIT subsidy approach might initially lead to more capacity than in the UK, the UK's competitive approach would lead to rapid price reductions, so that ultimately investment would flow and capacity would expand. The reality seems to have been different. Under the RO system, there is no certainty as to the future value of the ROC's, which makes it hard to get investment capital at reasonable interest rates for new projects, and those that have gone ahead had to charge higher prices to balance the uncertainty about future income. As we shall see, the prices that had to be charged for electricity have generally been higher than those under the REFIT system, and although the RO system kept the cost to consumers relatively low (it has led to an annual average of around a 0.5% increase), this was partly because relatively few projects went ahead.

The situation was made even worse by the UK governments introduction in 2001 of highly competitive New Electricity Trading Arrangements, which forced prices down across the whole electricity sector. That made it even harder for new renewable energy projects to get financed. Progress on the deployment of new projects slowed, with wholesale prices at an all time low, new projects required higher prices than could be provided by the RO.

By contrast, in Germany, and in many of the 17 other EU countries with REFIT-type schemes, not only was capacity expanding rapidly, but, as the market expanded, average prices fell to below those under the RO. For example, one comparison of EU wind projects showed that in 2003 the RO was delivering at 9.6 euro cents/kWh, while in Germany the REFIT scheme (or rather the revised version adopted in the 2000 Energy Law) was running at 6.6–8.8 euro cents/kWh. The REFIT-type schemes in the Netherlands, France, Portugal, Austria and Greece were also running at less than the RO. And the figure for Spain, which is fast becoming a wind leader, was 6.4 euro cents/kWh (Grotz & Fouquet, 2005). Overall, it has been calculated that the subsidy per kW of capacity installed provided by the RO so far may actually have been around 30% higher than that delivered by REFIT in Germany (Toke, 2004) and another study has suggested that the imbalance was likely to remain the case over the medium term (Butler and Neuhoff, 2004).

The UK's poor performance has been compounded by planning disputes in relation to wind farms, some of which can also be traced back to the competitive market approach. The competitive approach has meant that companies have located projects in more profitable high wind-speed upland areas – where they were often perceived as being more environmentally invasive. This has been one reason for the local backlash against wind projects – which has slowed deployment dramatically (Elliott, 2003).

By contrast, in Denmark, Germany and the Netherlands, projects have gone ahead using much lower wind speeds – indeed there was often no choice, since for example Denmark is mostly flat. In addition, as Toke describes in Chapter 8, many of these projects are locally owned – 80% of Denmarks' wind projects are owned by local farmers or community-based wind co-ops. And this has arguably been a key reason why there has been very little local opposition. As the Danish proverb goes 'your own pigs don't smell'. In the UK there are only two locally owned projects so far – the UK investment climate makes it hard for them to get established. However, some more have been trying – e.g. using funds raised by public subscription – and when and if community owned projects can get start up funding, then they should actually do well, since they presumably will face less local opposition and can then avail themselves of RO support, which ironically, is rather gen-erous (Toke, 2005). Even so, it is a uphill struggle for them to get going.

The RO also slowed another area of development which has prospered elsewhere – that is the creation of consumer-led demand for green power. Green power tariff schemes of various types have been offered to consumers across the EU and elsewhere and by 2006 over 3 million EU consumers had signed up to them. In some cases, this has provided a strong incentive for renewable energy development to meet the demand, as well as an extra cash income, given that most of the schemes involved a premium payment. However, in the UK context, the RO rules required that any power sold direct to consumers under such schemes had to come from sources *outside* the RO system, and there were not many of them. Moreover, what was available tended to be marginal and expensive. In the governments view, in terms of stimulating demand for new capacity, the RO was seen as the main way ahead, and the supply companies have inevitably focused on meeting their RO requirement first. Although most of the main supply companies have also offered green tariff schemes at various points, most of them, with possibly the exception of npower's non-premium 'Juice' scheme, have not tended to promote the schemes heavily, leaving this field mainly to a few independent energy retail companies. This is one reason why the so-called 'voluntary' green power market has not developed significantly in the UK – only around 200,000 consumers had subscribed by 2006. By contrast, in the Netherlands for example, over 40% of the total households in the country have signed up to green power schemes (Elliott, 2006).

## EU comparisons

As has been indicated, there are a number of reasons why, for the moment, the UK trails well behind most of the rest of the EU in terms of developing renewable energy capacity – many of them arguably being directly or indirectly linked to the RO. The results are certainly clear. Denmark has built up its wind programme to the level where it supplies around 20% of its annual average electricity requirements, whereas the figure for the UK only around 0.5%. This is very embarrassing, given that the UK has by far the best wind resource in the EU. The UK seems to have a system that is inflexible and ineffective, and one of the lowest levels of renewable energy production in the EU.

A report by the European Environment Agency has compared the successes and failures of EU renewable energy programmes between 1993–99. It noted that three countries that guaranteed purchase prices of wind-generated electricity via REFIT-type schemes – Germany,

***Table 9.1*** **Renewable energy as a percentage of total primary energy in 2001**

| | |
|---|---|
| Sweden | 29.4 |
| Finland | 22.4 |
| Austria | 21.5 |
| Portugal | 13.7 |
| Denmark | 10.4 |
| France | 7.0 |
| Spain | 6.5 |
| Italy | 5.6 |
| Greece | 4.6 |
| Germany | 2.6 |
| Ireland | 1.7 |
| Luxembourg | 1.6 |
| Netherlands | 1.4 |
| United Kingdom | 1.1 |
| Belgium | 1.0 |

*Source*: International Energy Agency / quoted in Hansard 15 Oct 2003: Columns 242.

Denmark and Spain – contributed 80% of new EU wind energy output during the period (EEA, 2002). By contrast, the UK was still trying to kick start its investment programme and build some capacity – the competitive market context was still unattractive to investors.

It is therefore hard to escape the conclusion that the competitive market-oriented renewable energy support system is one reason for the UK's relatively poor position in the EU renewable energy 'league table': see Table 9.1.

To be fair, one reason why countries like Sweden, Finland and Austria do so well in these league tables is because, unlike the UK, they have extensive and widely used biomass (forestry) resources, used mainly for heating, and also large hydro electricity resources.

It might be expected that this historical difference should gradually disappear, since the UK has the best wind, wave and tidal resources in the EU. However, the UK's 2010 target for electricity (thus excluding at least some of the biomass), is low compared with the targets agreed for most other EU countries, including some of the countries who joined the EU in 2004; see Tables 9.2 and 9.3. Furthermore, the situation does not improve much for the UK even when hydro is excluded (Table 9.2 right hand column). Basically, the UK seems not to be expected to achieve a major contribution. Worse still, the poor results, in terms of capacity achieved, from the RO system, may mean that even this relatively low level target will not be met.

***Table 9.2*** **EU directive: 2010 targets for electricity from renewables (% including and excluding hydro) (ranked in order of % excluding hydro)**

| | Including large hydro | Excluding large hydro |
|---|---|---|
| Denmark | 29.0 | 29.0 |
| Finland | 35.0 | 21.7 |
| Portugal | 45.6 | 21.5 |
| Austria | 78. 1 | 21.1 |
| Spain | 29.4 | 17.5 |
| Sweden | 60.0 | 15.7 |
| Italy | 25.0 | 14.9 |
| Greece | 20.1 | 14.5 |
| Netherlands | 12.0 | 12.0 |
| Ireland | 13.2 | 11.7 |
| Germany | 12.5 | 10.3 |
| UK | 10 | 9.3 |
| France | 21.0 | 8.9 |
| Belgium | 6.0 | 5.8 |
| Luxembourg | 5.7 | 5.7 |
| EU 15 | 22.1 | 12.5 |

***Table 9.3*** **EU accession country renewable electricity production in 1999 and % targets for 2010**

| Country | 1999 (%) | 2010 (%) |
|---|---|---|
| Cyprus | 0.05 | 6.0 |
| Czech Republic | 3.8 | 8.0 |
| Estonia | 0.2 | 5.1 |
| Hungary | 0.7 | 3.6 |
| Latvia | 42.4 | 49.3 |
| Lithuania | 3.3 | 7.0 |
| Malta | 0.0 | 5.0 |
| Poland | 1.6 | 7.5 |
| Slovakia | 17.9 | 31.0 |
| Slovenia | 29.9 | 33.6 |
| Average | 5.4 | 11.1 |
| EU 25 (target reduced from the EU15 target) | 12.9 | 21.0 |

*Source*: EC/Platts Renewable Energy Report I.

By contrast, as noted above, many of the renewable leaders, in terms of, for example, the rapid expansion of wind power, have all used, or are using, REFIT-type support systems – notably Denmark, Germany and, Spain. France has also now adopted a REFIT-type scheme, as has

Ireland. In addition, some of the new EU members have adopted REFIT-styled approaches. As Table 9.3 shows, Latvia currently gets a very large contribution from renewables, which it is planning to expand using a REFIT-style support system. Estonia has also adopted a REFIT type approach in its effort to develop its renewable energy resources very rapidly – by a factor of 5 by 2010.

In all by 2005, 18 of the EU's 25 countries were using REFIT-type schemes. The UK, Sweden, Italy and Poland had competitive quota/certificate trading-type schemes of various types, while Denmark, which initially used REFIT, was in the process of setting up a quota/certificate trading-type scheme. Slovakia and Lithuania use other types of support (Bechberger and Reiche, 2005).

Certainly many observers in the EU have seen REFIT as the best option for the present in terms of capacity development, developing a renewable power industry and bringing down cost and prices (Lauber, 2005). Outside of the EU, the situation is less clear. Wind power has developed quite rapidly in the US under state-based competitive quota systems, backed up by a federal production tax credit system. However the feed in tariff idea has been taken up in Canada, in the province of Ontario. It was also reflected in the support system initially proposed in China, as part of its new Renewable Energy Law. However, subsequently the approach has been changed – although there is a national fund scheme to provide grants for some key projects, and green tariffs have been introduced, a competitive tendering mechanism has also been introduced. It will be interesting to see how that plays out, given China's commitment to the rapid expansion of renewables to 16% of its significantly expanding electricity requirements by 2020 (Junfeng *et al.*, 2006).

## New technologies and innovation

While the advantages seem clear, REFIT-type systems also have their problems, for example in terms of loading up utilities with extra costs, some of which are then passed on to consumers. It has been estimated that the extra cost of the REFIT/'EEG system in Germany was around £1.28 billion in 2003. Although this is higher than the cost of the RO (the RO was expected to have raised around £500 million between 2003–06), it was for a very much larger capacity, and there seems to have been little concern about the resultant extra charge amongst German consumers. This is perhaps not surprising given that, in 2006, despite the relatively large number of projects that had been supported, the German Federal Environment Ministry claimed that the system only put around 1.6 euros extra on the average consumers monthly electricity bill, and it has been calculated that with the subsidy costing around 0.56 euro cent/kWh, it has added around 3% to household electricity costs (Stern, 2007).

More critically, it might be argued that the guaranteed price REFIT approach would inhibit innovation. Certainly just throwing money at projects may not always be a good idea. However, competitive pressures are not absent with REFIT, not least since, under REFIT-type schemes, after a period, the initial premium price is reduced in stages as the technology matures – a process known as 'degression'. Degression not only keeps overall prices down, it also ensures that there is pressure to reduce them by improving the technology and its operation, with the price reductions being set in line with the expected improvements indicated from the learning curves.

Clearly, generators who can make use of cost effective equipment will be better placed commercially than those that do not. Certainly, despite the REFIT-type schemes in Germany and, at one stage, Denmark, there seems to have been no lack of technological innovation, quite the opposite. Both these countries have been at the forefront of wind technology innovation, and both have large and prosperous wind turbine manufacturing industries. The same can hardly be said of the UK, which has no significant wind energy manufacturing industry – it imports the machines used for wind farms.

The same pattern has occurred in relation to PV solar. Germany in particular has pushed ahead with a generous support structure, well ahead of the market – PV is still relatively expensive. In relation to PV, and indeed the other renewables, the rationale for the REFIT system (and its subsequent replacement, the 'EEG') was that providing an initial subsidy would help create a large future market and a strong role for German industry in that market both at home and overseas. At that point the subsidy could be removed.

What about the future? Wind and solar are not the only renewables, and other options, for example biomass, wave and tidal power, are gradually coming on stream, and the UK is well placed to exploit at least some of these. However, here again the RO presents problems, while the REFIT approach has advantages. Under the RO, all projects compete equally against each other, regardless of the technology and its level of development. Which means that only the near-market options get picked up. But within REFIT type schemes, different prices are set for each type of renewable, reflecting their stage of development with, as noted above, the prices then being reduced (degressed) in planned stages to reflect expected technical and market developments. For example currently the level of support for wind is being reduced in Germany, while that for solar and biomass is still high – to help them move down their learning curves.

The advantages of this more flexible arrangement are clear – the system can give the necessary support to help new technologies get going. In the UK in 2005, the government began a review of the RO, but it insisted that it could only consider minor adjustments, so as to avoid disruption. It has however been offering special grants to try to push new technologies forward. So, in addition to capital grants for offshore wind projects and biomass projects, and a grant funding programme for PV solar, the UK now has a £42 million capital grants and revenue support scheme for wave and tidal projects, offering, for eligible new wave and tidal projects, grants of 25% of the capital cost, plus up to 10 p/kWh, *on top of* the support provided by the RO. You could see this as an admission that the competitive RO approach is not delivering – it has to be augmented with state subsidies and grants for new technologies (DTI, 2005a).

At the same time, as the National Audit Office (NAO) pointed out in a review of the RO in 2005, some of the technologies getting support from the RO have developed sufficiently to no longer need it – there being no degression mechanism. By 2010 it was expected that the RO would have raised around £1 billion, but the NAO noted that, *'the level of support provided by the RO is greater than necessary to ensure that most onshore windfarms and large landfill gas projects are developed'*. Moreover, it calculated that, unless the RO system was changed, by 2026–27, which is when the scheme was set to run to, *'around a third of the total public support provided could be in excess of that needed by generators to meet the higher costs of renewable generation.'* (NAO, 2005).

Faced with the prospect that some projects would be getting excess support, and keen to ensure that the RO capacity targets be achieved without putting extra cost on consumers, the government said it would consider withdrawing RO support from some of the more developed projects, in stages – landfill gas being one initial candidate. In effect it would be adopting some aspects of the REFIT system (DTI, 2005b).

However this may not be enough. In a review of the RO in 2006, the Carbon Trust concluded that the RO would not be sufficient to reach the UK targets unless it was improved much more radically, and it proposed a Renewable Development Premium 'market pull through' mechanism, similar to REFIT (Carbon Trust, 2006).

In the event, in its 2006 Energy Review, the government indicated that it would consider introducing 'technology bands' in the RO – something that it had earlier resisted. Each technology would be given a price level reflecting its state of development (DTI, 2006a). The details of this new approach, which is not seen as coming into force until 2009/10, have yet to emerge, but it does seem to be a recognition that the RO needed to be more like REFIT, at least for the newer renewables. As the DTI's RO reform consultation paper produced in October 2006 admitted '*As a technology-neutral instrument, the Renewables Obligation has thus far proved less successful in bringing forward development of the more emerging renewable technologies*' (DTI, 2006b).

The Energy Review also proposed that the annual level of the RO be adjusted so as to ensure there was always 'headroom' between current capacity levels and the targets, up to a 20% overall target. The aim was to ensure that the price of the Renewable Obligation Certificates (ROCs) earned by suppliers in the RO system would not fall as supplies reached the target that had originally been fixed at a 15% contribution to electricity by 2015. The DTI's October 2006 consultation paper proposed a 'ski slope' mechanism for ROC prices so that '*any renewable generation exceeding the level of demand for ROCs created by the Obligation would not have a precipitate impact on ROC prices but would instead ensure that ROC prices tapered smoothly down in a situation of oversupply*'. If accepted, this market intervention would certainly improve investor confidence in the RO system.

Even so, it will take time to make changes like this (the ski slope idea would require new primary legislation), and the UK is only likely to catch up slowly with countries who have already adopted the REFIT model. By mid-2006, Germany had around 18,000 MW of wind capacity in place, while the UK only had around 1,300 MW – with part of that being due to capital grants for offshore wind, provided *in addition* to the RO. By early 2007, the UK had moved ahead to 2 GW, but Germany had expanded to 20 GW.

In defence of the RO, the NAO's consultants did claim that the internal rates of return for wind project investments in the UK were similar to those elsewhere (NAO, 2005). However, you would expect at least that, given the UK's very much higher average wind speeds. What this really means is that, if a company can raise the money to get an RO contract, then they can make relatively large returns since, with relatively high prices, the RO can be relatively excessive in its level of support, especially for high wind speed projects which may not really need it.

This situation seems to have further antagonised people hostile to the UK wind power programme, some of whom, like the Renewable Energy Foundation and Ramblers Association, have called for the RO to be revamped to reduce support for wind projects, in favour of other types renewable energy and energy efficiency. However, it may also explain why the UK wind industry has in general been reasonably content with the RO system and has expressed concerns about any sudden changes, which it felt could disrupt it. Thus, although it has pressed continually for improvements in the planning system, and for the adoption of more ambitious overall targets so as to ensure that the value of ROCs would be protected, the British Wind Energy Association has argued that '*The RO is a market-based mechanism which has been very successful in incentivising new investors in the renewables sector*' adding that '*It is in no one's interest for development of the most cost effective renewable energy technology to be hindered by policy changes*' (BWEA, 2006).

It is certainly true that the UK renewables programme could do without any major disruptions, especially since a large number of wind projects have been held up by planning problems, but it is perhaps harder to accept the BWEA's claim that, in effect, the RO isn't broken so we don't need to fix it (EWEA, 2006). Even Windpower Monthly, a trade magazine that has backed the RO over the years, while still emphasising that there were many projects coming forward, felt moved to note that '*the government's aim with the Renewables Obligation was to secure as much green power as possible for as little money as possible. Part of that aim is being fulfilled: a record volume of new wind power capacity is coming online. But far from paying as little as possible, consumers in Britain probably have the highest priced wind power in the world*' (WPM, 2006).

## The EU single market and harmonisation

While some changes in the RO do now seem inevitable, the UK government did have a wider strategic justification for the RO system – with its quota and ROC trading element, it fits in with the competitive market approach that it, and the European Commission, would like to see adopted EU-wide as part of the drive to open up the EU energy market and harmonise the various support structures. In January 2005, the EC introduced an Emission Trading System (EU-ETS), a 'cap and trade' arrangement similar to the RO, but based on a carbon cap and trading carbon credits, rather than energy quotas and tradable renewable energy certificates (ROC's) as in the RO. In time, the ETS is seen by the EU as being one of the main mechanisms for driving renewables forward.

Certainly, in the long term, the existing renewables should not need REFIT-type subsidy systems, and should be able to operate in a competitive market environment which reflects the value of the emissions they help avoid, of the sort created by the EU-ETS. For the moment however, the EU has had to accept that the various national REFIT-type systems will continue in parallel – they have after all been so successful.

For example, in 2005 in its second progress report on the EU Renewables Directive, the European Commission admitted that Feed-in tariffs were '*currently in general cheaper and more effective than so-called quota systems, especially in the case of wind energy,*' and it accepted that it would be 'premature' to attempt to impose a single 'harmonised' renewable energy support scheme at this point. It even claimed that having a variety of national schemes '*can be healthy in a transitional period, as more experience needs to be gained,*' although this was still seen as interim stage, before a single EU-wide scheme could be selected (ReFocus, 2006).

Clearly the EC still saw harmonisation as the long-term aim: '*The integration of renewable energies in the internal market with one basic set of rules could create economies of scale needed for a flourishing and more competitive renewable electricity industry*', and it saw REFIT schemes as problematic. Although it *said 'These schemes have the advantages of investment security, the possibility of fine tuning and the promotion of mid- and long-term technologies'*, it argued that '*on the other hand, they are difficult to harmonise at EU level, may be challenged under internal market principles and involve a risk of over-funding, if the learning-curve for each RES-E technology is not build in as a form of degression over time*' (CEC, 2006).

However it admitted that '*harmonisation through a green certificate scheme with no differentiation by technology would negatively influence dynamic efficiency. Since such a scheme would promote cost-efficiency first, only the currently most competitive technologies would expand. While such an outcome would be beneficial in the short run, investment in other promising technologies might not be sufficiently stimulated through the green certificate scheme. Other policies would thus need to complement such a scheme*'. Just as the UK has found with the RO. And, interestingly, the EC said that '*a European wide common feed-in scheme which takes into account the availability of local resources could drive down the costs of all RES technologies in the different Member States*' (CEC, 2006).

For the moment at least that is not on the agenda, but the various existing REFIT type schemes in use in the EU are safe, leaving the UK, with the RO, and the four other countries using similar quota/trading arrangements, struggling with what seems to be an expensive way to develop renewables very slowly.

## Conclusion

The Renewables Obligation may have avoided adding significantly to electricity prices, but so far it has done this at the expense of building up capacity and has led to some high cost schemes, with higher overall prices than under the REFIT approach. The Stern review of climate policy for the UK Treasury summed up the situation as follows: '*Comparisons between deployment support through tradable quotas and feed-in tariff price support suggest that feed-in mechanisms achieve larger deployment at lower costs*', although it noted that '*greater deployment increases the total cost in terms of the premium paid by Consumers.*' In addition it commented: '*Contrary to criticisms of the feed-in tariff, analysis suggests that competition is greater than in the UK Renewable Obligation Certificate scheme. These benefits are logical as the technologies are already prone to considerable price uncertainties and the price uncertainty of tradable deployment support mechanisms amplifies this uncertainty. Uncertainty discourages investment and increases the cost of capital as the risks associated with the uncertain rewards require greater rewards*'. (Stern, 2007).

As noted above, to try to improve the situation, in the 2006 Energy Review, the UK government proposed some significant adjustments to the RO, including technology bands with different prices. In effect it seems to be trying to create something like a REFIT-type system, with a price differential and perhaps even a degression mechanism, while still retaining ROC trading.

The DTI seems keen on a 'multiple ROC' approach, e.g. offering double ROCs to newly emerging projects and partial ROCs to more developed options. However these changes will taken time to negotiate. Meanwhile, the Scottish Executive, is seeking to move ahead more rapidly, and is planning to set up a unilateral Scottish scheme – a Marine Supply Obligation with special funding just for wave and tidal projects (Scottish Executive, 2006). Whether these various adjustments and additions to the RO are the best way forward at this stage remains to be seen. It would have, arguably, been better to have adopted REFIT across the board from the start.

## References

Bechberger, M. and Reiche, D. (2005) 'Europe banks on fixed tariffs', *New Energy*, No. 2 April.

Butler, L. and Neuhoff, K. (2004) 'Comparison of Feed in Tariff, Quota and Auction Mechanisms to Support Wind Power Development', Cambridge University, Dept. of Applied Economics, Dec.

BWEA (2006) BWEA response to the Energy Review, July.

Carbon Trust (2006) 'Policy frameworks for renewables', London.

CEC (2006) 'The support of electricity from renewable energy sources' COM (2005) 627 final, SEC(2005) 1571, Commission of the European Communities, Brussels.

DTI (2005a) Marine Renewables: Wave and Tidal-stream Energy Demonstration Scheme, Department of Trade and Industry, London, Jan.

DTI (2005b) Statutory Consultation on changes to the Renewables Obligation, Department of Trade and Industry, London Sept. www.dti.gov.uk/renewables/renew_2.2.5.htm

DTI (2006a) 'The Energy Challenge' Government report of the Energy Review, Department of Trade and Industry, London, July.

DTI (2006b) Renewable Energy: Reform of the Renewables Obligation and Statutory Consultation on the Renewables Obligation Order 2007, Department of Trade and Industry, London, October.

EEA (2001) 'Renewable energies: success stories', Environmental issue report No. 27: Ecotec Research and Consulting Ltd. and A. Mourelatou, for the European Environment Agency, Copenhagen.

Elliott, D. (2003) 'Energy, Society and Environment', Routledge, London.

Elliott, D. (2006) 'Diffusion: consumers and innovation' Block 4: T307 'Innovation: Designing for a sustainable future', Open University, Milton Keynes.

EWEA (2006) Maria McCafferty BWEA CEO, interviewed in *Wind Directions*, Vol. 25, No. 5, European Wind Energy Association, July/Aug, pp. 20–2.

Grotz, C. and Fouquet, D. (2005) 'Fixed prices work better', *New Energy*, No. 2, April. Also see Grotz & Fouquet (2005), *Fixed prices work better*, German Wind Energy Association (BWE), downloadable from http://www.wind-energie.de/index.php?id=166

Junfeng, L., Jinli, S. and Lingjuan, M. (2006) 'China: Prospect for Renewable Energy Development' paper commissioned for the UK Treasury Stern Review: http://www.hm-treasury.gov.uk/independent_reviews/stern_review_economics_climate_change/stern_review_supporting_documents.cfm

Lauber, V. (ed) (2005) 'Switching to Renewable Power', Earthscan, London.

NAO (2005) 'Department of Trade and Industry: Renewable Energy', National Audit Office, London, Feb.

PIU (2002) The Energy Review, Performance and Innovation Unit, Cabinet Office, London.

ReFocus (2006) 'Europe rejects single green power scheme for continent', Refocus Weekly, March 1.

Scottish Executive (2006) Renewables Obligation (Scotland) – Statutory Consultation
http://www.scotland.gov.uk/Topics/Business-Industry/infrastructure/19185/ROSConsWaveTidal06Paper

Stern, N. (2007) 'The Economics of Climate Change' Review for HM Treasury by Sir Nicholas Stern, Cambridge University Press. http://www.hm-treasury.gov.uk/independent_reviews/stern_review_economics_climate_change/sternreview_index.cfm

Toke, D. (2004) 'Are Green electricity certificates the way forward for renewable energy?', Paper to 4th International Conference on Business and Sustainable Performance', Aalborg, Denmark, 14–15 April.

Toke, D. (2005) 'Don't write off community wind power', *Renew* 158 Nov–Dec, pp. 20–1.

WPM (2006) Windpower Monthly, On-Line Focus article, Oct.

# 10
# Carbon Trading: Opportunities and Issues

*Stephen Peake*

## Introduction

The creation of markets for carbon trading is often seen as a key way in which the development of new 'cleaner' energy options, as well as more environmentally appropriate industrial and corporate practices, might be stimulated.

At the heart of the global market in carbon trading is the Kyoto Protocol. The text of the treaty was finalised in late 1997, and it entered into force (i.e. became international law) in early 2005. Despite the recent rapid growth in the volume of transactions in the carbon market and the growing awareness and interest in its functioning on the part of politicians and the public, it is still very much in its infancy. Nevertheless, in a relatively short time, Kyoto has spawned a potpourri of linked or copy cat schemes ranging from internationally recognised and officially verified credits and allowances to locally generated voluntary and unverified credits.

This chapter provides an overview of the emerging carbon market in terms of the different schemes and carbon currencies of which it is comprised. It briefly examines the rationale for market-based carbon instruments, the history of allowance and credit schemes and then outlines some of the issues being raised by its critics.

## The ultimate value of avoided carbon emissions

It is beyond the scope of this chapter to describe in any detail how and why greenhouse gas concentrations are rising. The main cause is a growing global population (now approaching the seven billion mark and growing at the rate of one billion every 10–15 years or so) using increasingly greater amounts of fossil fuels and causing increasing amounts of land use change in their demand for energy, food and water.

The world is globalising but remains significantly divided between rich and poor. Roughly half the world's emissions of greenhouse gases that are regulated by the Kyoto Protocol are emitted by the wealthy developed economies. The other half is emitted by the developing economies – including the rapidly developing economies of China and India. However emissions from developing economies are growing extremely rapidly and very soon (within ten years or so) will overtake those from the developed economies.

Under the first Kyoto agreement agreed in 1997, the developed world agreed to cut its emissions of carbon dioxide, methane, nitrous oxide, perfluorocarbons, hydrofluoroucarbons and sulphurnexafluoride by around 5% relative to their 1990 levels by the end of 2012. This first Kyoto agreement does not include developing countries, though it was designed so that it could do so by amendment in subsequent 'commitment periods' – half decades marching on into the future 2012–17, 2017–22, 2022–27 etc.

The key point is that the political forces that shaped the Kyoto agreement were strongly in favour of what were called 'flexibility mechanisms'. There are three of them in the Kyoto agreement: emissions trading, the clean development mechanism and joint implementation. These mechanisms allowed signatories to the Protocol to meet part of their emissions reductions obligations by either trading emissions permits or investing in projects overseas where in greenhouse gas mitigation terms you get 'more bang for your buck'.

## The carbon market place

There are two fundamentally different sorts of schemes operating in the carbon market: *allowance*-schemes and project-based *credit* schemes.

Emissions trading schemes are allowances schemes. They are also sometimes referred to as a 'cap and trade' schemes. An allowance is an officially sanctioned amount of pollution – a permit to pollute. Organisations involved in emission trading schemes must ensure their emissions remain below their allocated allowance or face financial penalties. In theory, the penalties are set at a level such that it is cheaper to comply (for example by investing in new technology). Emissions trading between Nations under the Kyoto Protocol (they trade what are called Assigned Amount Units or 'AAUs') and the EU Emissions Trading Scheme are examples of such allowance schemes.

In project-based credit schemes an emissions credit is issued when a project is verified as having resulted in a reduction in carbon emissions. Reductions in greenhouse gas emissions are what is actually credited. Kyoto's two project-based mechanisms, the Clean Development Mechanism and Joint Implementation, are both project-based *credit* schemes (see below). In these schemes an emissions credit is a certificate issued when a project is verified as having resulted in the reduction in carbon emissions. Credits can be made up front before the project is completed (*ex ante*) or can be given only after the event (*ex post*).

Today's emerging carbon market – it is probably more accurate to say *markets* as they are not all linked – consists of a mixture of allowance and credit-based schemes. Some are directly related to the Kyoto Protocol, some tentatively related and others not at all. The overall market is becoming reasonably complex. For our purposes we can distinguish between:

- Kyoto related credits/allowances (from the CDM, JI, and AAUs)
- Other mandatory emissions trading schemes (e.g. the EU ETS, US State Cap and Trade schemes etc.)
- Various voluntary schemes (e.g. internal company emission trading schemes, retail carbon offset schemes).

To the financial markets, the different types of allowances and carbon credits differentiate in terms of 'quality' and in turn of course, market prices. Price Waterhouse Coopers, for example, distinguishes three categories:

- compliance carbon allowances and credits (these are any carbon allowances or credits that can be used for formal compliance within any regulatory regime. Examples include any credits recognised under Article 17 of the Kyoto Protocol (dealing with emissions trading), and credits from the CDM an JI
- pre-compliance carbon credits – quantifiable reductions from investments and projects which have the *potential* to become compliance carbon but have not yet met all the requirements (e.g. they may be waiting official verification or formal analysis of their baseline methodology, or a future political decision as to eligibility)
- voluntary (sometimes called retail) carbon credits – emission reductions from quantifiable and verifiable projects that are not designed to meet all the requirements for compliance with official regulatory regimes (Price Waterhouse Coopers, 2003).

The market value of any emissions credit or allowance reflects the amount someone or some entity is willing to pay for it. In the regulated market, it is the penalty or fine that someone faces for non compliance with any scheme they are in that ultimately underpins prices. In the voluntary sector it is the 'feel good' or marketing value (to a company) of demonstrating one's green credentials.

While true Kyoto compliance credits trade at premium prices, their supply is limited in the short term and dependent on several factors ultimately governed by political decisions in the international climate change negotiations. The potential market in retail credits is however unlimited by the international political process. PWC point out that 'all in all a properly structured carbon retail market could deliver substantially greater emissions reductions than those driven by Kyoto compliance.' This chapter will discuss some of the issues raised by the growth in interest in voluntary carbon offsetting.

## Kyoto Protocol-Related carbon instruments

The complex architecture of the Kyoto Protocol comprises what are known as the three 'flexibility mechanisms': emissions trading; the Clean Development Mechanism (CDM); and joint implementation (JI). They are flexible in the sense that they offer developed nations the chance to offset their domestic (national) emissions by funding and/or purchasing carbon 'offsets' abroad. The economic case for the flexibility mechanisms is that they are market-based and as such offer all the usual advantages over command and control approaches to environmental regulation. In theory, the flexibility mechanisms allow emission reductions to be made where they are cheapest. Critics argue that they are a distraction from getting on with the more fundamental and politically painful task of tackling emissions 'at home'. The flexibility mechanisms were a key component in the original 1997 negotiations between developed and developing countries on the draft text for the Kyoto Protocol.

### *Emissions trading*

The economic ideal of emissions trading makes great sense. The economic logic flows as follows:

- The cost of reducing emissions varies from one person, organisation, company, technology, project, nation to another
- The market knows best where to find cheap reductions

- If nations/organisations are free to find cost-effective emission reductions and avoid expensive ones, then the overall cost of compliance with environmental targets will be lower.

The two alleged principle benefits of emissions trading above traditional command and control regulation are increased cost-effectiveness and increased speed of compliance. International emissions trading under the Kyoto Protocol (known technically in trading circles as 'Article 17 trading' – after the relevant clause in the text of the Protocol) allows a list of around 40 countries (comprising developed OECD nations as well as economies in transition) to trade with each other as a way of fulfilling their national emissions quotas (technically when a Nations trade in carbon under Article 17 they are exchanging allowance units – AAUs).

Emissions trading was first introduced in the US (Tietenberg, 2006). The emissions trading and carbon project elements of the Kyoto protocol were strongly supported by the US in the 1997 Kyoto negotiations. By that time, the US had tried various forms of allowance- (and credit-) based environmental trading schemes. The term offsetting in the sense of emissions has its origins in attempts by the US Environmental Protection Agency (EPA) to encourage compliance with the Clean Air Act. In the late 1970s, the EPA encouraged emitters in so-called 'non attainment areas' to voluntarily reduce their emissions below legal requirements. These 'offsets' were then certified by the EPA and offered to new sources/emitters entering the non attainment areas. This was known as the Offset Policy.

A short time later the offset policy was expanded to become the US Emissions Trading Program which allowed existing sources to acquire credits from each other and also to bank credits for the future. This first emissions trading programme was a credit program. Credits were created by firms reducing their emissions below requirements and having the difference certified as an emissions reduction credit. Later, the US Acid Rain Programme introduced allowances to utility plants to emit sulphur oxides. Sulphur oxide emissions had been legally capped and emissions could not legally exceed allowances. There are many detailed innovations in trading that occurred in the US – the main point is that this experience informed and shaped the text of the Kyoto Protocol and therein much of the design of the mechanisms elaborated under it.

### *The European Emissions Trading Scheme*

Separate from, but linked to, the Kyoto Protocol is the European Union's Greenhouse Gas Emission Trading Scheme (EU ETS). The first phase of the scheme ran from 2005–07. The second phase of the scheme runs from 2008–12 and includes all six greenhouse gases regulated under Kyoto.

The scheme includes over 12,000 installations, representing approximately 45% of EU $CO_2$ emissions. Sectors covered include energy production (combustion installations with a rated thermal input exceeding 20 MW, mineral oil refineries, coke ovens), production and processing of ferrous metals, minerals (cement clinker, glass and ceramic bricks) and pulp, paper and board making.

The EU ETS is an allowance scheme where allowances are given freely to Nations (who in turn pass them onto the installations affected) who submit a National Allocation Plan.

Outside of Kyoto-based emissions trading, the EU ETS is do date the largest and most advanced emissions trading scheme in the world. Other countries are considering similar schemes (e.g. Canada, Japan, New Zealand).

*Project-based mechanisms under the Kyoto Protocol*

Through the CDM, developed countries can claim emission reduction credits from projects undertaken in developing countries – again to exploit the cost-effectiveness of reducing emissions where costs are lower, while at the same time to promote sustainable development. The name given to credits under the CDM is 'certified emission reductions' (CERs).

Joint Implementation is the name given to projects undertaken by the developed countries in other developed countries and countries with economies in transition. The idea is essentially to exploit the opportunities for low cost emissions reductions in the former Soviet countries. The name given to JI credits is Emission Reduction Units (ERUs).

CDM projects accounted for 95% of the volume of project transactions under the Kyoto Protocol (just under 400 MtC in 2005 according to PointCarbon). JI is tiny in comparison – around 30 MtC.

The notion that CDM and JI (and indeed voluntary) climate projects have ancillary benefits (e.g. contributing towards sustainable development goals) is a key feature and rationale of most climate projects. The degree to which these theoretical benefits are realised in practice is the source of much debate and has resulted in criticism from environmental NGOs such as Carbon Trade Watch, Sinks Watch, FERN and many others.

Officially recognised CDM projects represent the 'gold standard' of all such climate projects. The scrutiny they face is supposed to be unparalleled anywhere else in the carbon market. However the effectiveness of the overall CDM as well as specific programmes and projects within it has been challenged from official scrutiny such as the UK's Stern Report. According to Stern:

> The CDM in its current form is making only a small difference to investment in long-lived energy and transport infrastructure. While a substantial international flow of funds is being generated through CDM, it falls significantly short of the scale and nature of incentives required to reduce future emissions in developing countries. (Stern, 2007, pp. 570–1).

According to the Guardian Newspaper, by the end of 2012 the CDM is expected to have resulted in 1.4 billion tonnes of $CO_2$ emission reductions from around 400 projects officially approved by the CDM's executive board (*Guardian*, 2006).

However, over half of this achievement is due to a roaring trade in HFC 23 reduction projects mainly in India and China. Relatively few CDM projects related to energy efficiency, fuel switching, biomass or other renewables have been certified.

## Consumer carbon offsets

Consumer carbon offsets are another example of emissions credits – but they remain totally unconnected to the official accounts of the Kyoto Protocol. Governments, individuals, corporations and even celebrities are all involved in the trade in carbon offsets supplied by an array of unregulated offset companies. Purchasers motives vary from the need to reduce future carbon liabilities or quite simply to assuage guilt about their carbon habits and dependency. Consumers in particular are being offered offsets in increasingly high profile ways as international companies such as British Airways take up the cause. Offset companies are doing a brisk trade out of consumer guilt around climate change.

Relevant academic literature on the 'voluntary' or 'retail' or 'consumer' carbon offset market is sparse. One relevant and useful policy report was published by the UK's DEFRA in 2001 (DEFRA, 2001) and provides a useful general framework to guide analysis of the topic. The report distinguishes different types of offsets covering several phases of the lifecycle of products and services.

With regard to the possibility of rules regarding offsets, the report notes the following:

- No clear guidelines are in place relating to carbon offset advertisements and the way in which messages relating to carbon offsets are communicated.
- That this type of distinction between carbon offsets could be combined with a requirement that all claims on the product are clear and consistent and make it clear to the consumer which phase of life-cycle has been offset.
- Feedback from four consumer focus groups suggested that there were differing and highly conflicting opinions on the issue of how offsets should be applied to products.
- When questioned consumers do express concerns about the 'quality' of carbon offsets (DEFRA, 2001: p. 4).

At the time of writing, the UK's DEFRA has not produced any further publicly available materials relating to carbon offset schemes since the 2001 report.

The authors of the report recommend:

- That the Government should at a minimum introduce some kind of quality assurance (QA) system for the source of offset project activity – the QA system would result in some kind of 'kite mark' being available for use on products and services that would provide assurance that genuine $CO_2$ reductions had been achieved. Use of the 'kite mark' could be subject to a minimum standard of information quality and quantity on the product about the offset.
- That the Government establish a new set of rules for carbon offset projects to allow for ex ante accounting and the inclusion of projects that are currently not permitted under the UK Emissions Trading Scheme.

The report suggests the following core guiding principles and categories:

- Offsets should not be linked to qualities inherent in the product itself (e.g. they should not be used to signal products which emit less carbon than others in the range).
- It is not appropriate to use consumer offsets to encourage people to buy low energy products.
- Labels and associated marketing material must give clear explanations.
- There is a balance to be stuck between stringency and market visibility.

- The possibility should be considered of offering partial offsets for organisations (DEFRA, 2001: p. 3).

Labelling rules or codes of practice for the offset business would need to consider a range of complex information. Disclosure of the characteristics of an offset in a 'Kite Mark' scheme could contain information on the following elements:

- Type of offset (*ex gratia*, in use, freestanding, organisational (partial or otherwise), lifecycle, manufacturing, whole life cycle).
- Type of project (renewable, energy efficiency, forest).
- Timescale (how long will it take for at least 50% of the claimed reduction effect to have take place – e.g. within the next five years, ten and 50 years).
- Verifiability (has the offset been certified as real, measurable, and additional (e.g. Kyoto compliant, non-Kyoto compliant).
- Traceability (Are reductions traceable – is there an accountable connection between the consumption of the product or service that is creating the offset and the physical aspects of the offset project itself (particular tree, or stand, light bulb, or organisation).
- Permanence (is it permanent or temporary and if so how temporary).
- Location of risk (with advertiser, offset company, climate (no allocation of risk essentially leaves the risk to the climate)
- Social and Environmental Sustainability (additional information on extent to which project meets any relevant criteria).

Clearly there is a considerable agenda to be addressed before voluntary carbon offsetting might be seen as a reliable approach to achieving emission reductions – something we will return to below. This may not be surprising given that consumer offset schemes have emerged outside the regulatory framework. By contrast, it might be expected that the formal schemes that derive directly from the Kyoto protocol and which are overseen and regulated by major institutions, should be more robust. However the realty is that both their operation and design have attracted some major criticisms.

## Issues with carbon trading

Many environmental NGOs and commentators have raised fundamental questions about the efficacy of the emerging carbon market.

Organisations such as Carbon Trade Watch have been a consistent critic of many aspects of emissions trading and project-based approaches to solving climate change over a number of years.

Criticism for the CDM has come from official quarters too. The Stern report says the project-based nature of the CDM market 'creates issues of moral hazard and gaming, where there are incentives to manipulate the system to increase the rewards received (or reduce the costs paid).' (Stern, 2007, p. 572)

Consumer carbon offset schemes have begun to attract significant negative publicity. In 2006 the New Internationalist ran a highly critical special issue on the subject (New Internationalist, 2006). Other negative coverage has included Ergo, 2005, *The Guardian*, 2006 and Monbiot, 2006.

Opposition to carbon trading has emerged from the following roots:

- Anti-capitalist, anti-free trade critics of neo-liberal environmental strategies.
- NGOs campaigning on forest (sinks) related issues – in particular the rights of local indigenous peoples.
- Commentators on consumption – critics of consumer-based carbon offset schemes.

Below we will look at some of the main criticisms.

*The anti-trade/environmental injustice perspective*

In 2003 Carbon Trade Watch published a report '*The Sky is Not the Limit: the emerging market in Greenhouse Gases*' produced by the Transnational Institute (TNI, 2003). The report is a stinging vitriolic attack on the idea of that 'people are being cheated in the name of sustainable development.'

Their opposition to carbon trading is multifaceted and reflects the broad range of other NGO constituencies that have also raised concerns. The report's conclusion are based on the following premises:

- Early forms of pollution trading in the US led to environmental injustice (the alleged creation of toxic hotspots as a result of US emissions trading schemes such as RECLAIM brought in during the mid-1990s).
- The World Bank's Prototype Carbon Fund promotes privatisation in poor countries.
- The arguments by corporate lobbyists around Kyoto and other international environmental agreements in favour of carbon trading use

coded language such as 'cost effectiveness' and 'efficiency' to transform regulatory threat 'into market opportunities by making themselves indispensable, diversifying risk, evading responsibility and by so doing ensuring institutional survival' (p. 13).

- The co-optation of NGOs in the process – 'Corporate culture is hypnotising environmental NGOs with multi-stakeholder dialogues. Big business has shaped itself into human form and become a "stakeholder" in society' (p. 15).
- Opposition to carbon sinks (forest) projects – 'Projects in countries such as Uganda and Ecuador have already led to thousands of local communities dependent on forest areas being forced off their land as private Northern corporations backed by their governments engage in a world wide land grab at wholesale prices' (p. 16).
- The lack of robustness in the rules surrounding Kyoto Trading Regime as well as the efficacy of some of the specific architecture (e.g. whether or not Kyoto mechanisms with benefit renewables).
- Carbon trading is simply an extension in the logic of free trade – 'Free trade in greenhouse gases' – links between Kyoto and Trade and Investment.

Carbon Trade Watch's general stance towards trade in carbon credits is neatly summed up in this paragraph:

> As soon as climate change is described as a problem of a scarce resource being used irrationally, much the way food and water are discussed in many international for a, two avenues of problem-solving immediately appear in the prevailing neo liberal politics of our times. First, the response to scarcity is to define property rights and protections for investors. This analysis can be recognised in statements like 'water is not unlimited and people will only value water if they have to pay for it'. When translated into everyday life this means replacing shared street free water for every house with individual pre-paid water meters as has been seen in the South African township of Orange Farm in 2002 (p. 13).

*Ethical – 'leadership' in the climate negotiations*

A fundamental complaint – and one that has been there right from the start of negotiations on Kyoto – relates to the issues of 'supplementarity' as it is known. This is the argument that developed nations should take the lead in mitigating greenhouse gases and that this should be done 'at home' with their own domestic emissions inventories. The text of the Kyoto Protocol recognises this as a valid concern and limits developed countries use of the mechanisms to be 'supplemental' to domestic action.

Referring to project-based mechanisms, Article 6, paragraph d states that 'the acquisition of emission reduction units shall be supplemental to domestic actions'. Referring to emissions trading, Article 17 states that 'any such trading shall be supplemental to domestic actions'.

Right from the start of Kyoto then we have a tension between any economic or other practical arguments for carbon trading (in whatever form) and political considerations of what is ethically permissible. This is rocket fuel for the critics.

*Projects involving carbon sinks*

The politics of sinks projects reflects the debate in the scientific literature on the role of sinks as a greenhouse gas mitigation option. This debated is structured around a number of scientific, policy relevant and political questions.

Scientific elements include:

- What is the magnitude of the terrestrial carbon sink, how has this been changing over time and what factors are responsible for changes in the amount of carbon stored on land? (e.g. Houghton, 2002).
- Debates on the nature of scientific evidence on the relative importance of (a) physiological or metabolic factors that affect rates of sequestration and (b) disturbance and recovery mechanisms including changes in land use management that affect the amount of carbon stored in forests. Scientific evidence now suggests that land management (past and present) is having a much larger effect than $CO_2$ fertilisation (e.g. Houghton, 2002).
- The feasibility of identifying and factoring out sinks resulting from mechanisms unrelated to direct human management so as to avoid undue credits and debits in the regime of international climate accounting (e.g. Houghton, 2002).
- The impact of forestry projects and other disturbances on soil carbon in different climatic conditions (e.g. Saleska *et al.*, 2003, Smith, 2004).
- Methodologies for full carbon accounting (e.g. Nilsson *et al.*, (2000).

Policy-relevant and political elements include:

- The role of forest projects in assisting with global and/or macro-economic, social and environmental objectives such as poverty alleviation, capacity building and overall economic development.
- The extent to which carbon credits could manage to reduce long-term trends in deforestation in developing countries.
- The relative cost-effectiveness of forest projects as an option for greenhouse gas mitigation.
- Whether or not project-by-project, multi-project or national inventory approaches are optimal policy instruments.
- The extent to which standardised procedures for forest verification can be developed.
- Optimal instruments or mechanisms to address the non-permanence of carbon offsets from forest projects (see Richards and Andersson, 2001).

A key issue in sinks project is verifiability. It is possible to verifiably quantify $CO_2$ offsets achieved by tree planting but within a margin of error. The margin of error depends on the specific circumstances of the project. Estimated offsets are not the same as actual verified offsets. *Ex ante* (before hand) verification and certification of emission offsets achieved by tree planting is subject to a variety of risks that could result in the partial reversal of those offsets. *Ex post* (after the event) verification can confirm actual carbon capture and storage but only within the project boundary. All carbon offset projects (forest or otherwise) require careful monitoring, *ex post* verification, and certification over the lifetime of the project. In the case of forest projects, the longer the guaranteed life of the project, the more 'permanent' those offsets become.

There is no doubt that accounting for carbon offset projects presents new methodological challenges. 'There is considerable uncertainty associated with the estimates derived using the techniques that will be required to monitor, quantify and verify land carbon sinks established under the Kyoto Protocol' (Royal Society, 2001: p. vii).

All sinks projects (CDM or voluntary consumer carbon offsets) rely on the same techniques for monitoring, quantifying and verification and therefore face the same uncertainties.

It has been pointed out that the question of whether or not it is possible to estimate with confidence and at reasonable cost the carbon effects of individual forest projects is not dissimilar from the methodological challenges faced in cost-benefit analysis, environmental impact statements and economic and project development assessments (Richards and Andersson, 2001).

A range of uncertainties effect carbon offset measurements at various scales including:

- At the local and project specific level – including for example: the size and nature of the project (whether or not it is an afforestation or reforestation project, the types of species involved etc), whether or not whole forest ecosystem effects are taken into account (e.g. impacts on soil carbon), the time frame (according to the IPCC TAR 'newly planted or regenerating forests, in the absence of major disturbances, will continue to uptake carbon for 20 to 50 years or more after establishment, depending on species and site conditions, though quantitative projections beyond a few decades are uncertain' (IPCC, 2000: para 10)).
- Baseline/additionality uncertainties – human and natural systems are dynamic, making it at best difficult to say what would have happened in the absence of the project (and also in many instances to say whether or not the project would have happened in the absence of the programme).
- Leakage: changes within the project boundary can effect emissions outside the project boundary (e.g. shifting of trade from rehabilitated areas to outside the project boundary).

The range of different types of uncertainty involved in any particular forest project tends to compound and cascade resulting in wide margins of error overall. There is a great deal of scientific debate on various aspects of the topic of carbon sinks and sequestration projects (Aukland *et al.*, 2003). We must be careful to distinguish scientific debate relating to:

- National emissions inventory methods for full carbon accounting – the establishment of monitoring, reporting and verification system under Articles 3.3 and 3.4 of the Kyoto Protocol. For example the UK provided official figures of it carbon pool size for 2,000 with a margin of ±30–60% depending on the particular land use inventory sub category (UK's 2004 inventory).
- Individual projects – A key point is that of cost – measurement uncertainty can be reduced if enough money is spent. The IPCC Special Report on Land Use Change summarises the situation as

follows: 'Techniques and tools exist to measure carbon stocks in project areas relatively precisely depending on the carbon pool. However, the same level of precision for the climate change mitigation effects of the project may not be achievable because of difficulties in establishing baselines and due to leakage. Currently, there are no guidelines as to the level of precision to which pools should be measured and monitored. Precision and cost of measuring and monitoring are related. Preliminary limited data on measured and monitored relevant aboveground and below-ground carbon pools to precision levels of about 10% of the mean at a cost of about US$1–5 per hectare and US$0.10–0.50 per ton of carbon have been reported. Qualified independent third-party verification could play an essential role in ensuring unbiased monitoring.' (IPCC, 2000; para 82).
- There is as yet no approved baseline methodology for reforestation or forestation projects in the CDM.

Carbon sinks can be estimated using a variety of methods, techniques and tools including:

- Direct measurements of vegetation and soil at local scales (standard forest inventory methods: estimates of stem volumes, total tree biomass). If a hectare of trees is planted, it is a fairly easy matter to come back every five years and measure how much they have grown. Their diameters are measured and then a species-specific allometric equation is used to turn diameter into volume of timber. From there, the density of the wood is used to find the mass of wood, and from there a carbon content of 50% is assumed. To scale up from foresters' measures of stem volume to total tree volume (including roots), there are published 'biomass expansion factors'. In practice, it isn't difficult uncertainty is in the range +/– 10%. Forest projects also affect levels of soil carbon, generally in the direction of an increase. It is more difficult to detect changes in soil carbon, because there is a lot of spatial heterogeneity. Most carbon projects assume there are no changes in soil carbon to reduce uncertainty in the baseline.
- Indirect measurements such as extrapolation using GIS and/or remote sensing techniques. There is of course a cost involved in having to visit sites and re-measure trees. Remote sensing by satellite hasn't really made it yet, except in the narrow sense of letting us check whether forests are still there. In the tropics this is really useful.

The Royal Society concluded that 'uncertainties associated with estimates derived from all these methods are considerable' and that 'a fundamental problem is that the magnitude of this uncertainty is unknown' (Royal Society, 2001, p. 12).

It is important to note that there are no internationally agreed procedures and mechanisms for monitoring, verification and certification of forest carbon offsets.

There is no scientific controversy that there is indeed a carbon cycle and that tree planting can achieve significant carbon capture and storage (either temporarily or in some instances in the longer term). In its Third Assessment Report, the IPCC concluded that 'In practice by the year 2010 mitigation in land use, land use change, and forestry activities can lead to significant mitigation of $CO_2$ emissions' (IPCC, 2001b; p. 304). The Royal Society has estimated that managed sinks could potentially meet 25% of the reductions in $CO_2$ projected to be required globally by 2050 (Royal Society, 2001; p. 23).

*Issues relating to energy efficiency offset projects*

A number of similar issues surround carbon credits from energy efficiency projects. In particular, there are limits to the extent to which it is possible to verifiably quantify the degree to which energy efficiency projects represent an improvement over 'business as usual'. It is possible but once again only within a margin of error. The margin of error is project specific. A high degree of confidence in the accuracy of a project specific 'business-as-usual' baseline combined with good system of *ex post* project verification will result in a lower margin of error.

In practice, various uncertainties undermine confidence in the true value of energy savings from energy efficiency projects. Sources of uncertainty include:

- the baseline methodology.
- project leakage.
- monitoring and verification.
- the rebound effect (see Chapter 7).
- uncertainties due to the wider effects of the project in energy markets where there is suppressed demand.

There is a considerable literature dating back at least two decades on the subject of the evaluation of energy efficiency policies, programmes and projects. Demand-side energy efficiency projects are costly to verify with confidence because of the complexities of consumer behaviour (e.g. Khazzoom, 1987; Lovins, 1988; Parfomak and Lae, 1996; Nadel *et al.*, 1993; Eto *et al.* 1996).

In the case of replacing traditional bulbs with CFLs, for example, there is no scientific controversy about the power saving and efficiency improvements of, for example, an 18 watt CFL bulb. 18 watts is 18 watts, and such a bulb replacing a higher wattage bulb will produce the stated power saving. The extent of the energy savings over time is however less certain due to consumer behaviour. The only thing that replacing a CFL guarantees is a power saving, not an energy saving.

However, there is a mature debate in the scientific community that is essentially an economic debate about the effects that efficiency improvements have on people's behaviour and patterns of energy consumption.

Elements of that debate include for example:

- the size of the rebound effect of energy efficiency interventions. The rebound effect is a phenomenon based on economic theory and long-term historical studies, but as with all economic observations its magnitude is a matter of considerable dispute (Herring, 2004: p. 237).
- Appropriate methodologies for standardised approaches to baselines for project-based mechanisms such as the CDM (Gustavsson *et al.*, 2000).
- the dynamics of technical change and the autonomous energy efficiency improvement.
- the reconciliation of results from top-down macro energy-economy models with those of bottom up engineering models.
- Global energy market-based leakage *via* the price lowering effects of energy efficiency projects on energy markets outside of the Kyoto scheme (e.g. Chomitz, 2002).

To be credible, any demand side energy efficiency consumer carbon offset scheme or projects (e.g. CFLs, while goods, transportation) must factor rebound effects into their accurate assessments of project baselines.

Globally or at the level of a whole national economy it is difficult to distinguish the effect of demand-side energy efficiency policies on overall demand. Many variables change simultaneously (activity levels, energy prices, regulations, policies and measures) that it is extremely difficult to disassociate one effect from another.

For critics of the efficacy of energy efficiency this difficulty amounts to a lack of proof that energy efficiency policies actually work. In response, energy efficiency advocates point out that without the increased efficiency overall consumption would have been even greater. The future in the absence of intervention is unknowable. At the level of a national economy, neither side can conclusively prove their own case, nor that the other side is wrong. In practice, some of the benefits of energy efficiency will be taken up in the form of increased consumption of energy services (light, heat, vehicle miles, etc) or the monetary savings spent on other energy intensive goods and services (e.g. air travel).

Another concern is the extent to which the benefits of such projects offset emissions from the burning of fossil fuels.

It is important to distinguish two different approaches to approaching this issue:

- Rebound and energy market effects of energy efficiency projects in specific project contexts.
- The climatic consequences of fossil fuel emissions now versus effects of emission reductions in the future.

In order to claim that a project offsets emissions from the burning of fossil fuels it is necessary to establish this. Under special circumstances this could be established. The project would need metering (to establish savings) and system wide monitoring to verify that a reduction in emissions as a result of the project actually takes place. This would incur relatively high costs and is unlikely to happen.

Methodologies to establish the additionality of energy efficiency projects have been approved by the CDM Executive Board. Where a project meets or exceeds the CDM's standards for the proof of additionality and where it displaces fossil fuel use then it would be fair to claim that the project offsets emissions form the burning of fossil fuels. In other words, in the case where energy efficiency improvements as a result of a project have resulted in emissions reductions that have been independently verified as being real (additional) and further have been directly linked with a measured reduction in the quantity of fossil fuel burned in a system then it would be fair to claim that the project offsets emission from the burning of fossil fuels.

Paragraphs 28 and 29 of the simplified modalities and procedures for small-scale CDM project activities consider the baseline method in the case of displacement of emissions from a system including fossil fuel-fired power stations. However such a method does not appear to take into account system wide energy effects in demand constrained systems.

## Issues with carbon offsetting

As noted earlier, consumer carbon offsetting is part of what is known as the voluntary carbon market – people and organisations volunteer to reduce or neutralise their carbon footprints over and above what their governments have pledge to do as part of their Kyoto commitments. Voluntary carbon offsets are entirely separate from official carbon credit or allowances schemes such as Kyoto emissions trading, the European Emissions Trading Scheme or the Clean Development Mechanism. A ton of $CO_2$ voluntarily reduced is close to worthless on the carbon market. In contrast Kyoto compliance credits are highly valued.

Carbon offset projects, schemes and companies vary enormously in their design, management, location, and time horizons. Currently, there is no way easy way for consumers to judge quality. There are no legal standards, nor is there even a voluntary labelling scheme to help the consumer compare and contrast different projects/schemes in a systematic way. Electrical appliances have A, B, C ratings – carbon offsets do not.

Moreover, the term 'offset' can cover very different meanings and on its own it not an adequate description of exactly what is being offset and how this is being achieved or when this will occur. In addition, the quality of an offset can vary from one scheme to another, within each category of offset and between categories.

Supporters of carbon offsets point out that nevertheless offsetting is better than nothing. But, in fact, there is an argument that nothing would actually be better. There are several arguments against carbon offsetting:

- It is an unregulated market.
- It is ethically questionable – offsetting shifts the burden of changing consumption and technology patterns to other people, other places and other generations.
- It is very hard to offset to high ethical and sustainability standards – finding real and permanent emission reductions that local people want is extraordinarily difficult – even for big players and national governments. The costs involved are often very significant.
- Some schemes may actually damage local ecosystems and communities – depending on how they are designed, owned and managed.

Critics argue that the danger is that carbon offsetting becomes just another way to 'carry on consuming'. It shifts the burden of living within our ethical limits to some other place, some other person, some other time. Fly to New York today – and the trail of gas you leave starts work on warming the planet immediately. On the other hand, the tree you ordered to be planted will take 50–100 years before it will have sucked back enough carbon to offset the kerosene you burned.

There is a need for research that examines carbon offsetting in relation to emerging literature on the psychology of consumption. On the one hand carbon offsetting engages individuals and empowers them. It offers the chance of early action and a smaller footprint on an individual level. On the other, it could be said to 'disconnect' the effects of consumption from the activity itself. At best it disconnects, at worst it may shift the problem or even further may create even worse impacts.

## Conclusion

In theory, carbon markets can contributes towards sustainability in the following ways:

1. They can encourage nations/organisations to find cost-effective emission reductions and avoid expensive ones, and in doing so reduce the overall cost of compliance with environmental targets.
2. They can offer a clear transactional message of the connection between activity and environmental impact.
3. Through international cooperation around development projects, they can encourage technology transfer, and sustainable development.
4. They can result in real emission reductions or capture for some time (e.g. up to 99 years).

Key arguments used by critics of various aspects of the carbon market include:

1. At a National level it delays early domestic action (the supplementarity issue). At the level of voluntary carbon offsets, at best, it maintains consumption, at worst it may actually encourage it.
2. Some emissions credits in the system are not necessarily real or permanent. In the voluntary sector it is currently impossible to purchase a truly permanent offset.
3. They shift the burden to future generations.

As noted earlier, the uncertainties become particularly worrying in the case of voluntary carbon offsets. There can be many different types of carbon offsets and carbon offset schemes with varying qualities in terms of a range of criteria. But there are currently no carbon accounting standards, rules or guidelines that apply to voluntary consumer carbon credit schemes. There are no specific labelling rules, guidelines or codes of practice relating to the labelling and marketing of voluntary consumer carbon offset schemes. Clearly the situation needs to be improved before voluntary offsets can be relied on as a viable approach.

This chapter has examined the effectiveness of various emissions trading/carbon offsetting schemes. The theory is simple: healthy well regulated carbon markets offers the chance for the international community and even individuals to mitigate more cost effectively by allowing finance to flow towards more cost-effective opportunities. In practice, it remains to be seen whether or not firm links between international carbon markets, voluntary carbon offset markets and the ultimate goal of the stabilisation of atmospheric concentrations will be made.

## References


Aukland, L., Costa, P.M. and Brown, S. (2003) 'A conceptual framework and its application for addressing leakage: the case of avoided deforestation', *Climate Policy* 3 (2): 123–36, June 2003.

Chomitz, K.M. (2002) 'Baseline, leakage and measurement issues: how do forestry and energy projects compare?', *Climate Policy* 2 (1): 35–49, May 2002.

DEFRA (2001) 'Consumer Carbon Offsets: A study into the principles for operating schemes in the UK market and the types of schemes that should be included', Final Report, September 2001. Prepared by Energy for Sustainable Development Limited.

ERGO (2004) 'Celebs' tree planting company investigation' Winter, p. 12.

Eto, J., Vine, E., Shown, L., Sonnenblick, R. and Payne, C. (1996) 'The Cost and Measured Performance of Utility Sponsored Energy Efficiency Programs', *The Energy Journal*, 17 (1): 31–51.

*The Guardian* (2006) 'Trading spaces in the sky', Wednesday November 15, 2006.

Gustavsson, L., Karjalainen, T., Marland, G. *et al.* (2000) 'Project-based greenhouse-gas accounting: guiding principles with a focus on baselines and additionality', *Energy Policy* 28 (13): 935–46, Nov. 2000.

Herring, H. (2004) 'Rebound Effect of Energy Conservation', *Encyclopedia of Energy*, 5: 237–44.

Houghton, R.A. (2002) 'Magnitude, distribution and causes of terrestrial carbon sinks and some implications for policy', *Climate Policy* 2 (1): 71–88, May 2002.

IPCC (2000) Special Report on Land use, Land Use Change and Forestry, Summary for Policymakers, Intergovernmental Panel on Climate Change.

IPCC (2001b) 'Climate Change: Mitigation. Contribution of Working Group III to the Third Assessment Report of the Intergovernmental Panel on Climate Change', Cambridge University Press.

Khazzoom, J. Daniel (1987) 'Energy Savings Resulting from the Adoption of More Efficient Appliances', *The Energy Journal*, 8 (4): 85–9.

Lovins, A., (1988) 'Energy Savings Resulting from the Adoption of More Efficient Appliances: Another View', *The Energy Journal*, 9 (2): 155–62.

Monbiot, G. 'Buying Complacency', *Guardian* 17th January 2006.

Nadel, S., Atkinson, B. and McMahon, J. (1993) 'A Review of US and Canadian Lighting Programs for the Residential, Commercial and Industrial Sectors', *Energy*' 18: 145–58.

New Internationalist (2006) 'Carbon Offsets' special issue, *New Internationalist* No. 391, July.

Nilsson, P. *et al.* (2000) 'Full Carbon Account for Russia', IIAS Interim Report IR-00-021.

Parfomak, P. and Lae, L. (1996), 'How Many Kilowatts are in a Negawatt? Verifying Ex Post Estimates of Utility Conservation Impacts at the Regional Level', *Energy Journal*, 17 (4): 59–87.

PWC (2003) 'Will markets recognise "Premium Carbon"? They should!: A more expansive approach to valuing carbon certificates', Price Waterhouse Coopers position paper, London.

Richards, K. and Andersson, K. (2001) 'The leaky sink: persistent obstacles to a forest carbon sequestration program based on individual projects', *Climate Policy* 1: 41–54.

Royal Society (2001) 'The role of land carbon sinks in mitigating global climate change', Royal Society policy document, London Oct.

Stern, N. (2007) 'The Economics of Climate Change' Review for HM Treasury by Sir Nicholas Stern, Cambridge University Press. http://www.hm-treasury.gov.uk/independent_reviews/stern_review_economics_climate_change/sternreview_index.cfm.

Smith, P. (2004) 'Monitoring and verification of soil carbon changes under Article 3.4 of the Kyoto Protocol', *Soil Use and Management* 20: 264–70 Suppl. S June 2004.

Saleska, S.R., Miller, S.D., Matross, D.M., *et al.* (2003) 'Carbon in amazon forests: Unexpected seasonal fluxes and disturbance-induced losses', *Science* 302 (5650): 1554–7, Nov 28 2003.

Tietenberg, T.H. (2006) 'Emissions Trading Principles and Practice', Resources for the Future, Washington DC.

TNI (2003) 'The sky is not the limit', Trans National Institute, Carbon Trade Watch group, Amsterdam. www.carbontradewatch.org.

UNFCCC (2002) United Nations Framework Convention on Climate Change http://cdm.unfccc.int/pac/howto/SmallScalePA/ssclistmeth.pdf.

van Kooten, G.C., Eagle, A.J., Manley, J. and Smolak, T. (2004) 'How costly are carbon offsets? A meta-analysis of carbon forest sinks', *Environmental Science and Policy* 7 (4): 239–51.

# Part III

# The Future

# 11

# Long-term, Renewables-Intensive World Energy Scenarios

*Godfrey Boyle*

## Introduction

Can renewable energy sources make a major contribution to the world's growing energy needs, and to reducing atmospheric carbon emissions, during the 21st century? A number of detailed energy scenarios produced in recent years suggest that this is the case, and that it is feasible, technologically and economically, to reduce world atmospheric $CO_2$ emissions by 60–80% by mid-century, in order to avoid dangerous climate change. In most of these scenarios, developed by a variety of respected inter-governmental, governmental and non-governmental organisations, the rapid deployment of renewables (and in some cases other low- or zero-carbon energy sources), coupled with major improvements in the efficiency of energy use, play a major role in enabling fossil fuel use, with its associated carbon emissions, to be largely phased-out by the end of the 21st century.

At national level, the feasibility of major carbon reductions and renewable energy deployment has been illustrated by several detailed energy studies. In Germany, for example, a scenario produced in 2004 for the Ministry of Environment (BMU, 2004) envisages renewable energy contributing some 50% of the nation's energy demand by 2050. Coupled with major improvements in energy efficiency, this enables $CO_2$ emissions to be reduced by 80%. And in Britain, the Royal Commission on Environmental Pollution's 2000 report (RCEP, 2000) included four energy scenarios for the UK in 2050. In all of them renewable energy and energy efficiency improvements make major contributions (in varying proportions) to achieving a 60% cut in $CO_2$ emissions. (For a more detailed comparison of these and other scenarios, see Boyle, 2007; IEA, 2005; Boyle, 2004).

This chapter will focus, however, on world level, long-term, renewables-intensive scenarios, although some contrasting scenarios featuring more modest renewables' contributions will also be described. It will summarise a selected sample of scenarios produced by the International Energy Agency (IEA), the World Energy Council (WEC), the Intergovernmental Panel on Climate Change (IPCC), the International Institute for Applied Systems Analysis (IIASA), the Shell International Petroleum Company and Greenpeace International.

## The Renewables-Intensive Global Energy Scenario

One of the first scenarios in recent decades to envisage a much more significant role for renewable energy at world level was produced by a group of scientists for the 1992 United Nations Conference on Environment and Development in Rio de Janeiro. Entitled the 'Renewables-Intensive Global Energy Scenario' (RIGES) it concluded that, despite substantial economic growth and a consequent rise in total global energy consumption, 'by the middle of the 21st century, renewable sources of energy could account for three fifths of the world's elec-

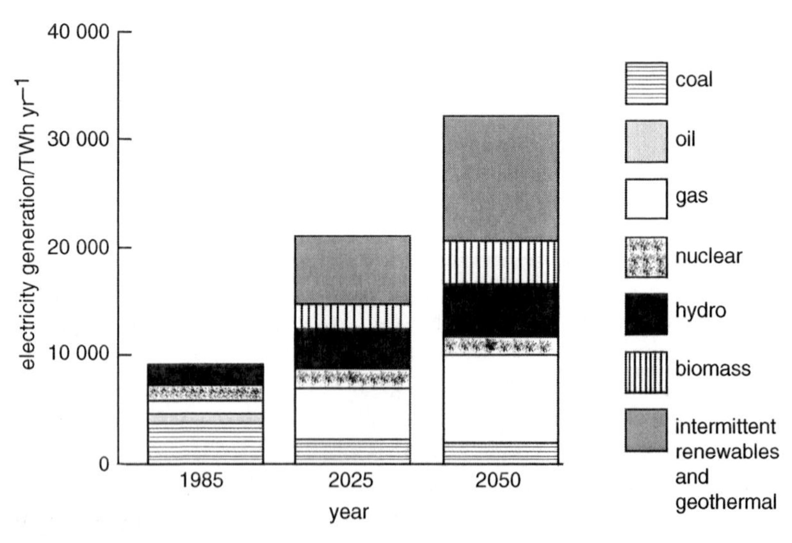


*Figure 11.1* Electricity generation in the Renewables-Intensive Global Energy scenario. The share of renewables grows to over 60% by 2050, despite a trebling of electricity consumption

*Source*: Johansson *et al*., 1993.

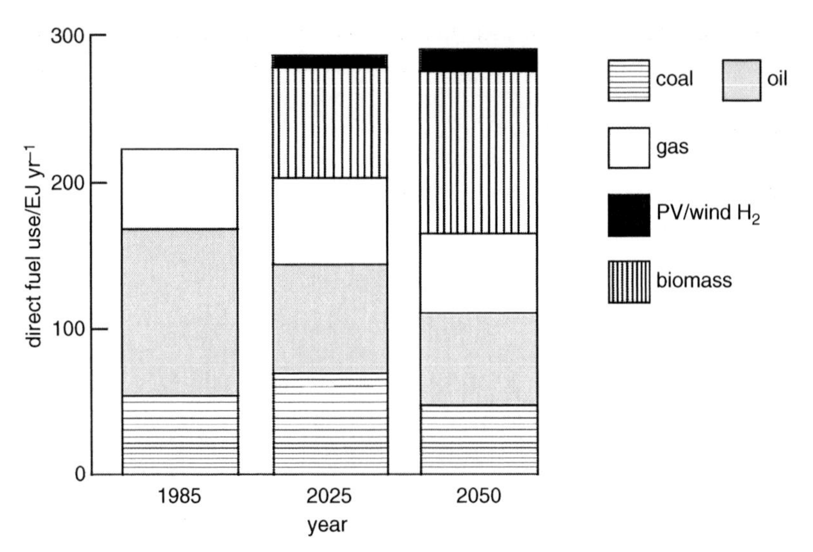


*Figure 11.2* Direct use of fuels for purposes other than electricity generation in the Renewables-Intensive Global Energy scenario. The share of renewables reaches 40% by 2050. ('PV/wind H2' is hydrogen generated from photovoltaics or wind power)

*Source*: Johannson *et al.*, 1993.

tricity market and two fifths of the market for fuels used directly.' See Figures 11.1 and 11.2.

## The World Energy Council scenarios

Later in the 1990s, a comprehensive set of long-term world energy scenarios for the 21st century was produced by the World Energy Council (WEC), in collaboration with the International Institute for Applied Systems Analysis (IIASA). The following account of them is based on the versions included in the United Nations *World Energy Assessment* (Goldemberg *et al.*, 2000) and Nakicenovic *et al.*, 1998.

The WEC/IIASA team produced a set of three basic 'families' of scenarios, called 'Cases'. The first, Case A, is 'High Growth' and includes three variants: A1, 'ample oil and gas'; A2, 'return to coal'; and A3, 'non-fossil future'. The second, Case B (which has no variants), is 'Middle Course' and represents a path to the future involving only gradual changes. The third, Case C, is 'Ecologically Driven' and includes two variants: C1, 'new renewables'; and C2, 'renewables and new nuclear'.

Each scenario incorporates different assumptions about rates of economic growth and the distribution of growth between rich and poor countries; about the choices that are made between different energy technologies and the rapidity with which they are developed; and regarding the extent to which ecological imperatives are given priority in coming decades. They all assume that world population will increase, in line with higher UN projections, from its 2000 level of around 6.1 billion to 10.1 billion by 2050 and 11.7 billion by 2100. (More recent projections (see Lutz *et al.*, 1997) suggest that these figures may be overestimates, however, and that world population is more likely to peak at around 9 billion around the middle of the 21st century and then begin to decline slowly. This point is discussed further below.)

The results of these assumptions are shown in Figure 11.3, which also shows (inset) world population growth from 1850–2000 alongside the various scenario projections to 2100. In all three High Growth scenarios, the world's economy expands very rapidly, at an annual average rate of 2.5% per annum – significantly faster than the historic growth rate of about 2% per year. In all of them, primary energy intensity (the amount of primary energy required to produce a dollar's worth of output in the economy) reduces quite rapidly, reflecting a fairly strong commitment to energy efficiency measures and/or 'dematerialisation' (see below).

By 2100, the High Growth scenarios all envisage world primary energy consumption rising to over 1,800 exajoules (EJ), more than three times the 2002 level (c. 450 EJ).

The three scenarios differ mainly in their choices of energy supply technologies. One is based on ample supplies of oil and gas; another envisages a return to coal; and the third has an emphasis on non fossil sources, mainly renewables with some nuclear.

In the single Middle Course scenario, economic growth is lower than in the High Growth scenarios, averaging around 2.1% per annum, close to the historic average rate. Primary energy intensity reduces rather more slowly, reflecting a slightly lower world-wide emphasis on energy efficiency improvement. Energy supplies come from a variety of fossil, nuclear and renewable sources, and by 2100 total primary energy consumption has reached more than 1,400 EJ, nearly three times the 2002 level.

In the two Ecologically-Driven scenarios, world economic growth is 2.2% per annum, slightly higher than in Middle Course, but there is a very high emphasis on improving energy efficiency, reflected in substantially lower primary energy intensity figures. Both scenarios feature a strong development of renewables, alongside a continued use of oil, coal and natural gas. In one scenario, nuclear energy is phased out by 2100 whereas in the other some nuclear power is retained. Overall primary energy consumption increases to some 880 EJ by 2100, just over twice the 2000 level.

The WEC/IIASA authors conclude that, judged in terms of their sustainability, one of the High Growth scenarios (the third) includes many elements favouring sustainable development, though the other two High Growth scenarios do not. The Middle Course scenario, however, falls short of fulfilling most of the conditions for sustainable development. The Ecologically-driven scenarios, unsurprisingly, are much more compatible with sustainable development criteria, although one of them requires a more radical departure from current policies since it envisages a phasing-out of nuclear energy.

The WEC's scenarios envisage the share of renewable energy rising rapidly, from around 22% by 2050 in the Middle Course scenario to between 22% and 30% in the High Growth scenarios, and up to 37–39% in the Ecologically Driven scenarios.

The three 'cases', and their variants, are illustrated in Figure 11.3 and the main scenario characteristics for 2050 are summarised in Table 11.1.

## The Shell scenarios, 1995 and 2001

In 1995, planners at Shell International Petroleum published the results of their detailed studies of the long-term future prospects for the world's energy system (see Shell International, 1995; Herkstroter, 1997). They created two exploratory scenarios sketching out two possible evolutionary paths for the world's energy system during the 21st century.

In the first of these scenarios, 'Sustained Growth', world energy demand increases to approximately 1,500 EJ by 2060, more than three times the 2002 level (c. 450 EJ), in the absence of any special efforts being made to introduce energy efficiency improvements. World GDP grows at 3% per annum and world energy demand at 2% per annum. By the middle of the next century, the world's energy sources are more diversified than at present, with new renewables supplying around 50% of global demand for commercially-traded energy. Fossil fuel use is beginning to decline, not so much because of absolute limitations on supply but because renewables have become more competitive in price.

In the second scenario, called 'Dematerialisation', the world economy becomes more frugal in its use of materials and energy. Technological developments converge to create breakthroughs in the efficiency of products and processes. Nevertheless, energy demand rises at around 1% per annum to around 1,000 EJ by 2060, with the renewables' con-

***Table 11.1*** **The World Energy Council scenarios: characteristics of the three 'Cases', and their variants, for the world in 2050 compared with 1990**

| **Base year:** | **1990** | **Case A** | | | **Case B** | **Case C** | |
|---|---|---|---|---|---|---|---|
| | | (A1) | (A2) | (A3) | | (C1) | (C2) |
| **Primary energy, Gtoe** | 9 | 25 | 25 | 25 | 20 | 14 | 14 |
| *Primary energy mix, percent* | | | | | | | |
| Coal | 24 | 15 | 32 | 9 | 21 | 11 | 10 |
| Oil | 34 | 32 | 19 | 18 | 20 | 19 | 18 |
| Gas | 19 | 19 | 22 | 32 | 23 | 27 | 23 |
| Nuclear | 5 | 12 | 4 | 11 | 14 | 4 | 12 |
| Renewables | 18 | 22 | 23 | 30 | 22 | 39 | 37 |
| *Resource use 1990 to 2050, Gtoe* | | | | | | | |
| Coal | | 206 | 273 | 158 | 194 | 125 | 123 |
| Oil | | 297 | 261 | 245 | 220 | 180 | 180 |
| Gas | | 211 | 211 | 253 | 196 | 181 | 171 |
| **Energy sector investment, trillion US$** | 0.2 | 0.8 | 1.2 | 0.9 | 0.8 | 0.5 | 0.5 |
| **US$/toe supplied** | 27 | 33 | 47 | 36 | 40 | 36 | 37 |
| **As a percentage of GWP** | 1.2 | 0.8 | 1.1 | 0.9 | 1.1 | 0.7 | 0.7 |
| **Final Energy, Gtoe** | 6 | 17 | 17 | 17 | 14 | 10 | 10 |
| *Final energy mix, percent* | | | | | | | |
| Solids | 30 | 16 | 19 | 19 | 23 | 20 | 20 |
| Liquids | 39 | 42 | 36 | 33 | 33 | 34 | 34 |
| Electricity | 13 | 17 | 18 | 18 | 17 | 18 | 17 |
| Other[a] | 18 | 25 | 27 | 31 | 28 | 29 | 29 |
| *Emissions* | | | | | | | |
| Sulfur, MtS | 59 | 54 | 64 | 45 | 55 | 22 | 22 |
| Net carbon, GtC[b] | 6 | 12 | 15 | 9 | 10 | 5 | 5 |

*Note*: Subtotals may not add due to independent rounding.
[a]District Heat, gas and hydrogen
[b]Net carbon emissions do not include feedstocks and other non-energy emissions or $CO_2$ used for enhanced oil recovery.
*Note*: GWP is Gross World Product.

*Source*: Nakicenovic *et al.*, 1998.

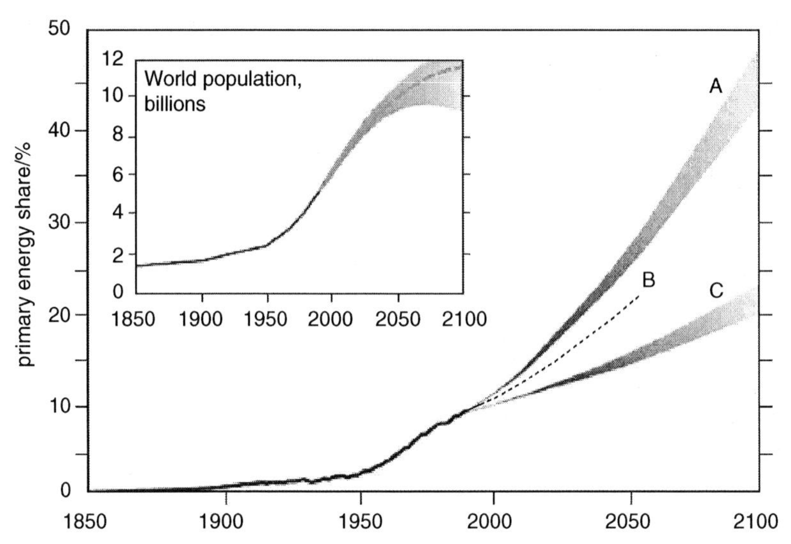


*Figure 11.3* Global primary energy use (in Gtoe): historical development from 1850 to 1990 and in the three World Energy Council 'cases' (scenarios) to 2100. Inset: global population growth 1850–1990 and projections to 2100
*Note*: 1 Gtoe is equivalent to 42 Exajoules (EJ).

*Sources*: Goldemberg *et al.*, 2000; Bos *et al.*, 1992.

tribution slightly lower than the 50% envisaged in 'Sustained Growth' because of the lower overall demand.

In 2001, Shell International went on to publish two new long-term energy scenarios: 'Dynamics as Usual' and 'Spirit of the Coming Age'. 'Dynamics as Usual' envisages 'an evolutionary progression from coal to oil to gas to renewables (and possibly nuclear).' But 'Spirit of the Coming Age' is more revolutionary, envisaging 'the potential for a hydrogen economy – supported by developments in fuel cells, advanced hydrocarbon technologies and carbon dioxide sequestration' (Shell International, 2001).

Figure 11.4 and Table 11.2 show Shell's analysis, for the 'Dynamics as Usual' scenario, of the market shares of various fuels in the world fuel mix for the past (1850–2000), and its projections of the likely future shares from 2000–50. (Note that the timescale here, ending in 2050, is slightly shorter than in the earlier Shell scenarios, which ended in 2060). These envisage overall world demand rising to some 850 EJ by 2050, with the share of coal, oil and traditional biofuels continuing to decline as the 21st century progresses. The share of nuclear power stays relatively constant, the share of gas increases to around

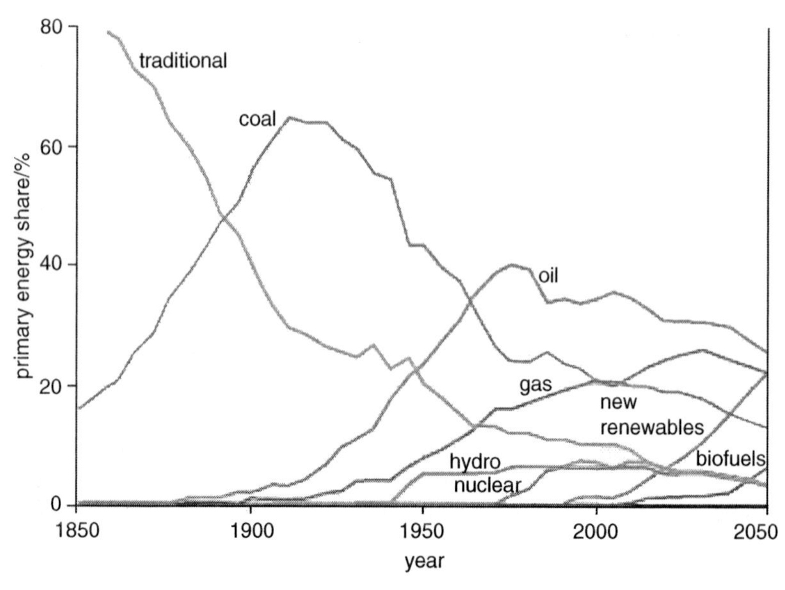


*Figure 11.4* In Shell's 'Dynamics as Usual' scenario, energy supplies continue to evolve from high- to low-carbon fuels and towards electricity (from increasingly distributed sources) as the dominant energy carrier, driven by demands for security, cleanliness and sustainability

*Source*: Shell, 2001.

2030 and then declines slightly, the share of hydro increases and then levels-out, and the share of modern biofuels and 'new' renewables steadily grows. By 2050, the combined market share of all renewables, including 'new' renewables, hydropower, new and 'traditional' biofuels, rises to around 33% by 2050.

Shell's projections for its 'Spirit of the Coming Age' scenario are shown in Figure 11.5 and Table 11.2. This scenario envisages world demand by 2050 rising to c. 1,120 EJ, a substantially higher level than in Dynamics as Usual. The share of solid and liquid fuels steadily declines, but the share of gaseous fuels, in the form of either natural gas or hydrogen, steadily grows, as does the share of electricity from hydro, nuclear and 'new' renewables. By 2050, the share of all renewables reaches around 30% of total demand.

In both of Shell's 2001 scenarios, the overall shares of renewable energy projected for 2050, at around one third, are significantly lower than the approximately 50% share projected in their 1995 predecessors. Shell's commitment to expansion of natural gas in the medium

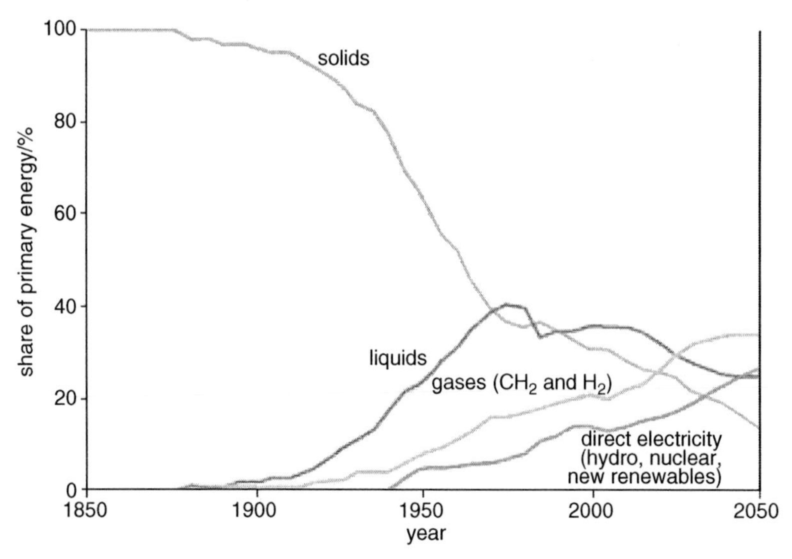


*Figure 11.5* In Shell's 'Spirit of the Coming Age' scenario, energy supplies continue to evolve from solids through liquids to gas (first methane then hydrogen), supplemented by direct electricity from renewables and nuclear, and by long-term sources of hydrogen gas

term may have played a key role in this re-evaluation of the relative roles of renewables and fossil fuels. As Philip Watts, chairman of the Shell group's committee of managing directors, observed in his introduction to the 2001 report: 'Expanding the use of gas is perhaps the most important immediate way of responding to the climate threat, as well as of improving air quality.' (Shell International, 2001).

## Intergovernmental panel on climate change (IPCC) 'SRES' scenarios

In 2000 the Intergovernmental Panel on Climate Change produced a *Special Report on Emission Scenarios* (SRES) (IPCC, 2000). It included six scenario 'story lines' and associated scenario 'families' depicting various possible paths that the world's energy systems and their associated emissions might follow over the 21st century.

The first storyline and scenario family, A1, envisages very rapid economic growth, with world population peaking at 8.7 billion around 2050 and then slowly declining to c. 7 billion by 2100; swift introduction

***Table 11.2*** **Summary of the main characteristics of Shell's 'Dynamics as Usual' and 'Spirit of the Coming Age' scenarios**

| Dynamics As Usual Scenario | 1975 | 2000 | 2025 | 2050 | Annual growth rate 1975–2000 | Annual growth rate 2000–25 | Annual growth rate 2025–50 |
|---|---|---|---|---|---|---|---|
| World Population (billion) | 4 | 6 | 8 | 9 | 1.50% | 1.00% | 0.60% |
| World GDP (trillion 2000 $ PPP*) | 23 | 49 | 108 | 196 | 3.10% | 3.20% | 2.40% |
| **Dynamics as Usual** | **1975** | **2000** | **2025** | **2050** | **Annual growth rate 1975–2000** | **Annual growth rate 2000–25** | **Annual growth rate 2025–50** |
| Primary Energy (EJ) | 256 | 407 | 640 | 852 | 1.90% | 1.80% | 1.20% |
| Oil (EJ) | 117 | 159 | 210 | 229 | 1.20% | 1.10% | 0.30% |
| Coal (EJ) | 70 | 93 | 128 | 118 | 1.10% | 1.30% | –0.30% |
| Coal CH4/H2# (EJ) | 0 | 0 | 4 | 16 | 0.00% | 0.00% | 5.90% |
| Natural Gas (EJ) | 47 | 93 | 167 | 177 | 2.70% | 2.40% | 0.20% |
| Nuclear (EJ) | 4 | 29 | 35 | 32 | 8.10% | 0.80% | –0.40% |
| Hydro (EJ) | 17 | 30 | 41 | 39 | 2.40% | 1.30% | –0.30% |
| Biofuels (EJ) | 0 | 0 | 5 | 52 | 0.00% | 10.20% | 10.10% |
| Other Renewables (EJ) | 0 | 4 | 50 | 191 | 8.70% | 11.20% | 5.50% |
| **Spirit of the Coming Age Scenario** | **1975** | **2000** | **2025** | **2050** | **Annual growth rate 1975–2000** | **Annual growth rate 2000–25** | **Annual growth rate 2025–50** |
| Primary Energy (EJ) | 256 | 407 | 750 | 1121 | 1.90% | 2.50% | 1.60% |
| Oil (EJ) | 117 | 159 | 233 | 185 | 1.20% | 1.60% | –0.90% |
| Coal (EJ) | 70 | 93 | 150 | 119 | 1.10% | 1.90% | –0.90% |
| Coal CH4/H2# (EJ) | 0 | 0 | 6 | 97 | 0.00% | 0.00% | 11.60% |
| Natural Gas (EJ) | 47 | 93 | 220 | 300 | 2.70% | 3.50% | 1.30% |
| Nuclear (EJ) | 4 | 29 | 46 | 84 | 8.10% | 1.90% | 2.40% |
| Hydro (EJ) | 17 | 30 | 49 | 64 | 2.40% | 2.00% | 1.10% |
| Biofuels (EJ) | 0 | 0 | 7 | 108 | 0.00% | 11.80% | 11.80% |
| Other Renewables (EJ) | 0 | 4 | 38 | 164 | 8.70% | 9.90% | 6.00% |

*Note*: Nuclear, hydro, wind solar and wave contributions are expressed as thermal equivalents.
* PPP denotes GDP estimate made on a Purchasing Power Party basis
# Denotes methane or hydrogen manufactured from coal

*Source*: Shell International, 2001.

of new and more efficient technologies; continuing globalisation; and an increasingly equitable distribution of income between regions.

The A1 family of scenarios has three variants. In the first, A1B, fossil fuel technologies are emphasised, leading to higher greenhouse gas emissions and greater risk of adverse climate change. The second variant, A1T, emphasises the development of non-fossil fuel technologies, leading to lower emissions and reduced climate change risk. The third variant, A1F1, includes a balanced mix of fuels and technologies, with more moderate reductions in emissions.

The single A2 storyline and scenario describes a more self-reliant world in which local identities are preserved, with slower economic growth and technological change, continuously increasing global population but slower growth in per-capita income.

In this scenario, emissions of greenhouse gases and other pollutants increase significantly.

In the B1 storyline and scenario, population growth is similar to that in the A1 Scenario (peaking around mid-century), and this is coupled with sustained economic growth in an increasingly-globalised world economy. Clean and resource-efficient technologies are rapidly introduced. These developments enable higher per-capita incomes and improved inter-regional equity to be achieved, together with major environmental improvements.

The B2 storyline and scenario describes a world in which local solutions are sought to problems of economic, social and environmental sustainability. The emphasis is on individual behavioural change rather than technological change. Population steadily increases, though at a slower rate than in the A2 scenario, and economic growth and technological change are less rapid than in other scenarios.

Across these six scenarios, the role of renewables in primary energy supply varies markedly. At one extreme, in the A1T variant, the contribution of renewables (including biomass) rises to some 79% by 2100; and in the A1B variant, they make a 61% contribution. In the middle are the B2 variant, in which the renewables share is 39%; and the A1F1 and B1 variants, each with a 20% share. At the other extreme is the A2 variant, in which the renewables contribution by 2100 is only 14%.

The main characteristics of SRES family of scenarios are summarised in Table 11.3.

## The IEA 'Sustainable Development (SD) Vision' scenario

The International Energy Agency publishes a range of medium-term energy projections, the best-known of which is probably its *World Energy Outlook*, published annually. The 2006 edition includes energy projections to 2030 (IEA, 2006).

*Table 11.3* Key characteristics of the IPCC 'SRES' family of scenarios

| | 1990 | | | | | | 2050 | | | | | | 2100 | | | | | |
|---|---|---|---|---|---|---|---|---|---|---|---|---|---|---|---|---|---|---|
| | A1-B | A1-F1 | A1-T | A2 | B1 | B2 | A1-B | A1-F1 | A1-T | A2 | B1 | B2 | A1-B | A1-F1 | A1-T | A2 | B1 | B2 |
| Population, Million | 5262 | 5293 | 5262 | 5282 | 5280 | 5262 | 8704 | 8703 | 8704 | 11296 | 8708 | 9367 | 7056 | 7137 | 7056 | 15068 | 7047 | 10414 |
| GNP/GDP (mer) Trillion US$ | 20.9 | 20.7 | 20.9 | 20.1 | 21.0 | 20.9 | 181.3 | 164.0 | 187.1 | 81.6 | 135.6 | 109.5 | 528.5 | 525.0 | 550.0 | 242.8 | 328.4 | 234.9 |
| GNP/GDP (ppp) Trillion (1990 prices) | | na | 26 | | 3971 | 26 | | na | 186 | | 15569 | | 114 | na | 535 | | 46598 | 232 |
| **Primary Energy (EJ)** | | | | | | | | | | | | | | | | | | |
| Coal | 93 | 88 | 91 | 92 | 105 | 91 | 186 | 475 | 119 | 294 | 167 | 86 | 84 | 607 | 25 | 904 | 44 | 300 |
| Oil | 143 | 131 | 128 | 134 | 129 | 128 | 214 | 283 | 250 | 228 | 228 | 227 | 125 | 248 | 77 | 0 | 99 | 52 |
| Gas | 73 | 70 | 71 | 71 | 62 | 71 | 465 | 398 | 324 | 275 | 173 | 297 | 576 | 578 | 196 | 331 | 103 | 336 |
| Nuclear | 6 | 24 | 7 | 8 | 8 | 7 | 123 | 137 | 115 | 62 | 105 | 48 | 78 | 233 | 114 | 234 | 165 | 142 |
| Biomass | 50 | 0 | 46 | 0 | 3 | 46 | 193 | 52 | 183 | 71 | 95 | 105 | 376 | 123 | 370 | 162 | 67 | 315 |
| Other Renewables | 10 | 24 | 8 | 8 | 61 | 8 | 167 | 86 | 222 | 42 | 46 | 107 | 987 | 284 | 1239 | 86 | 36 | 212 |
| Total | 376 | 336 | 352 | 313 | 368 | 352 | 1347 | 1431 | 1213 | 971 | 813 | 869 | 2226 | 2073 | 2021 | 1717 | 514 | 1357 |
| **Primary Energy %** | | | | | | | | | | | | | | | | | | |
| Coal | 25 | 26 | 26 | 29 | 28 | 26 | 14 | 33 | 10 | 30 | 21 | 10 | 4 | 29 | 1 | 53 | 8 | 22 |
| Oil | 38 | 39 | 37 | 43 | 35 | 37 | 16 | 20 | 21 | 23 | 28 | 26 | 6 | 12 | 4 | 0 | 19 | 4 |
| Gas | 19 | 21 | 20 | 23 | 17 | 20 | 35 | 28 | 27 | 28 | 21 | 34 | 26 | 28 | 10 | 19 | 20 | 25 |
| Nuclear | 2 | 7 | 2 | 3 | 2 | 2 | 9 | 10 | 9 | 6 | 13 | 5 | 3 | 11 | 6 | 14 | 32 | 10 |
| Biomass | 13 | 0 | 13 | 0 | 1 | 13 | 14 | 4 | 15 | 7 | 12 | 12 | 17 | 6 | 18 | 9 | 13 | 23 |
| Other Renewables | 3 | 7 | 2 | 3 | 17 | 2 | 12 | 6 | 18 | 4 | 6 | 12 | 44 | 14 | 61 | 5 | 7 | 16 |
| Total | 100 | 100 | 100 | 100 | 100 | 100 | 100 | 100 | 100 | 100 | 100 | 100 | 100 | 100 | 100 | 100 | 100 | 100 |
| **Cumulative Resources Use, ZJ** | | | | | | | | | | | | | | | | | | |
| Coal | 0.1 | 0.1 | 0.0 | 0.0 | 0.0 | 0.0 | 9.1 | 14.6 | 8.1 | 9.5 | 8.5 | 5.7 | 15.9 | 37.9 | 11.7 | 46.8 | 13.2 | 12.6 |
| Oil | 0.1 | 0.1 | 0.0 | 0.0 | 0.0 | 0.0 | 12.7 | 10.4 | 11.3 | 13.9 | 11.6 | 12.0 | 20.8 | 29.6 | 20.8 | 17.2 | 19.6 | 19.5 |
| Gas | 0.1 | 0.1 | 0.0 | 0.0 | 0.0 | 0.0 | 13.9 | 12.7 | 9.7 | 8.7 | 7.5 | 8.6 | 42.2 | 40.9 | 25.0 | 24.6 | 14.7 | 26.9 |
| Cumulative CO2 Emissions GtC | 0.0 | 0.0 | 0.0 | 0.0 | 0.0 | 0.0 | 730.6 | 820.9 | 623.1 | 728.6 | 599.0 | 554.5 | 1492.1 | 2182.3 | 1061.3 | 1855.3 | 975.9 | 1156.7 |

*Source*: http://sres.clesin.org/final_data.html

In a 2003 report, *Energy to 2050: Scenarios for a Sustainable Future,* the Agency examined a range of longer-term energy scenarios for the middle of the 21st century. These included a 'Sustainable Development (SD) Vision' scenario.

The IEA 'SD Vision' scenario is based on the IPCC A1T scenario (see Table 11.3 above) and is a 'normative' one, explicitly designed to achieve a 60% share of 'zero carbon' sources in the world's primary energy supply by 2050. 'Zero carbon' here includes both renewables and nuclear power – and, importantly, fossil fuel technologies with carbon capture and storage.

Compared with the IPCC A1T scenario, SD Vision incorporates more proactive policies to encourage the use of zero carbon fuels. As can be seen from Table 11.4, the share of 'zero carbon' sources (i.e. renewables, including biomass, and nuclear power) is c. 46% by 2050, compared with 42% in the A1T scenario.

To achieve the desired 60% reduction in emissions, the SD Vision scenario envisages the capture and long-term storage of some 26% of all emissions from fossil fuel combustion. Key characteristics of the SD Vision scenario are summarised in Table 11.4 and in Figure 11.6.

## The IIASA 'Post-Fossil' and 'Oil and Gas-Rich' scenarios

In 2005, IIASA published detailed descriptions of two contrasting energy scenarios for the 21st century, respectively entitled 'Oil-Gas Rich' (OG) and 'Post-Fossil' (PF), produced with the aid of IIASA's 'MESSAGE' modelling system (Schrattenholzer *et al.*, 2005), used in the creation of its earlier scenarios.

The OG and PF scenarios are both based on the same assumptions regarding population growth and economic growth. World population is projected to peak at around 8.7 billion by 2060, and then decline to c. 7.1 billion by 2100 (Lutz, 1996, 1997). As mentioned above, this more recent projection is somewhat more 'optimistic' regarding the prospects for limiting future population increases than the projection by Bos *et al.* used in the WEC and some other scenarios.

World GDP is projected to rise from $21 trillion in 1990 to $550 trillion in 2100 – a 26-fold increase. The authors believe that 'policy-making in general is made easier when economic growth is fast. (and that)... high economic growth also means the introduction of new (and the replacement of old) energy-generating equipment...'. More generally, and contrary to some other authors, they believe that sustainable development and economic growth can be compatible.

*Table 11.4* Characteristics of the SD Vision scenario

| | 1990 | 2000 | 2010 | 2020 | 2030 | 2040 | 2050 |
|---|---|---|---|---|---|---|---|
| Population (million) | 5262 | 6117 | 6888 | 7617 | 8182 | 8531 | 8704 |
| GNP/GDP (ppp) trillion (1990 prices) | 26 | 33.4 | 43.0 | 64.9 | 94.9 | 131.2 | 173.2 |
| Primary Energy – EJ | | | | | | | |
| Coal | 91.10 | 105.6 | 118.3 | 135.6 | 153.5 | 128.0 | 99.3 |
| Oil | 128.3 | 155.0 | 165.8 | 178.5 | 193.0 | 191.4 | 181.3 |
| Gas | 70.5 | 86.9 | 123.2 | 157.3 | 206.9 | 244.2 | 267.1 |
| Nuclear | 7.3 | 8.2 | 11.4 | 18.1 | 39.1 | 75.3 | 114.5 |
| Biomass | 46 | 45.5 | 52.8 | 69.3 | 92.3 | 117.5 | 159.0 |
| Other renewables | 8.3 | 14.8 | 25.1 | 45.6 | 71.6 | 122.1 | 191.8 |
| Total | 351.5 | 416.0 | 496.6 | 604.3 | 756.3 | 878.5 | 1013.0 |
| Energy for transport – EJ | 66.7 | 90.5 | 119.7 | 155.8 | 217.7 | 277.7 | 344.4 |
| Oil – EJ | 64.0 | 80.4 | 98.4 | 111.3 | 130.6 | 131.2 | 132.4 |
| | **1990** | **2000** | **2010** | **2020** | **2030** | **2040** | **2050** |
| Primary Energy – % shares | | | | | | | |
| Coal | 25.9 | 25.4 | 23.8 | 22.4 | 20.3 | 14.6 | 9.8 |
| Oil | 36.5 | 37.3 | 33.4 | 29.5 | 25.5 | 21.8 | 17.9 |
| Gas | 20.1 | 20.9 | 24.8 | 26.0 | 27.4 | 27.8 | 26.4 |
| Nuclear | 2.1 | 2.0 | 2.3 | 3.0 | 5.2 | 8.6 | 11.3 |
| Biomass | 13.1 | 10.9 | 10.6 | 11.5 | 12.2 | 13.4 | 15.7 |
| Other renewables | 2.4 | 3.6 | 5.1 | 7.5 | 9.5 | 13.9 | 18.9 |
| Total | 100.0 | 100.0 | 100.0 | 100.0 | 100.0 | 100.0 | 100.0 |
| of which zero-carbon | 17.5 | 16.5 | 18.0 | 22.0 | 26.8 | 35.8 | 45.9 |
| Transport share of primary energy % | 19.0 | 21.8 | 24.1 | 25.8 | 28.8 | 31.6 | 34.0 |
| Oil as a % of total transport energy | 95.9 | 88.8 | 82.2 | 71.4 | 60.0 | 47.2 | 38.4 |
| Oil for transport in total oil % | 49.9 | 51.9 | 59.4 | 62.4 | 67.7 | 68.6 | 73.0 |
| Emissions from fossil fuels GtC | 5.78 | 6.90 | 7.98 | 9.19 | 10.68 | 10.57 | 9.99 |

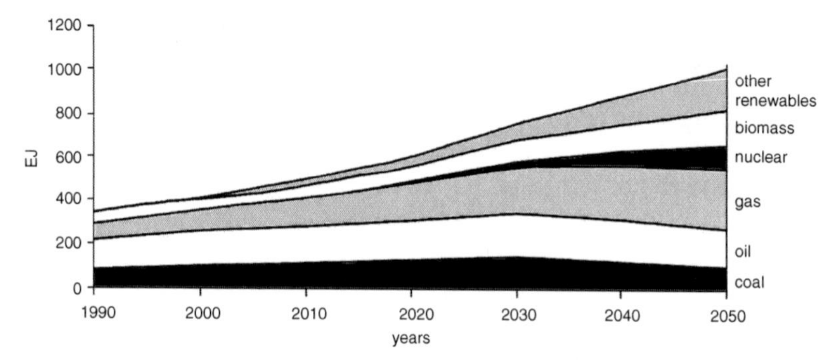


*Figure 11.6* Primary energy supply mix, 1990–2100, in the IEA 'SD Visions' scenario

*Source*: IEA, 2003.

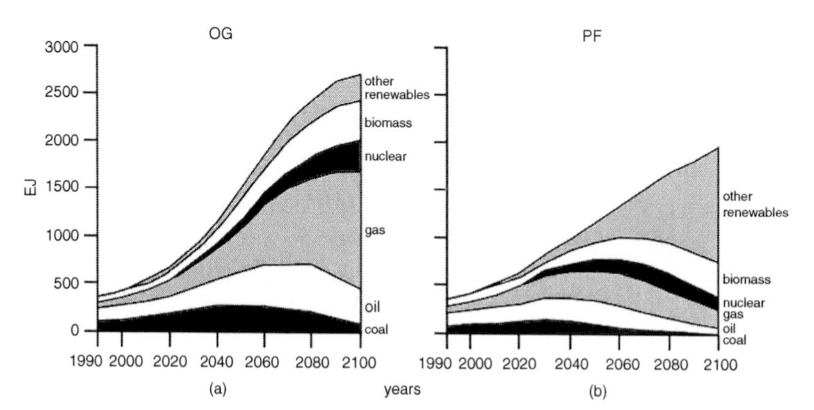


*Figure 11.7a and 11.7b* The IIASA Oil and Gas Rich (OG) and Post-Fossil (PF) scenarios

*Source*: Schrattenholzer *et al.*, 2005.

Inequality between rich and poor regions of the world, as measured by income per capita, is projected to decrease markedly in both scenarios, from a ratio of 1:16 in 1990 to 1:1.6 in 2100.

The Oil and Gas Rich (OG) scenario describes a future '... in which technological change is concentrated on the oil and natural gas sectors (...) reflected in cost reductions in unconventional oil and gas extraction and conversion technology, and substantial improvements and extensions of the present pipeline grids...' In the OG scenario, as shown in Figure 11.7a, world primary energy demand rises from c. 350 EJ in 1990 to c. 2,700 EJ in 2100, with renewables (including biomass) contributing approximately 25%.

In the Post-Fossil (PF) scenario, by contrast, 'technological progress is concentrated on conversion technologies fuelled mainly by renewable energy, on technologies that produce or utilise synthetic fuels including hydrogen, as well as on efficiency improvements of end-use technologies.'

Primary energy demand in the PF scenario, as shown in Figure 11.7b, rises to c. 2,000 EJ by 2100, considerably lower than in the OG scenario, with biomass and other renewables contributing the majority of primary energy – around 80%.

The two scenarios differ dramatically in terms of atmospheric carbon emissions and the resulting atmospheric $CO_2$ concentrations. In the OG, scenario, global carbon emissions rise from c. 6 GtC/year in 1990 to around 30 GtC per year by 2100, a level likely to result in an atmospheric carbon concentration of c. 900 ppmv and a global mean temperature rise of c. 2.8°C.

In the PF scenario, on the other hand, global carbon emissions rise to c. 14 GtC/year by around 2040, but then fall to around 4 GtC/year by 2100, probably leading to an atmospheric $CO_2$ concentration of c. 550 ppmv and a global mean temperature rise of c. 1.9°C – just under the 2.0°C rise considered by many authorities to be a critical level below which global temperatures should be stabilised if 'dangerous climate change' is to be avoided (see Schellnhuber *et al.*, 2006). For this and other reasons, the IIASA authors consider their PF scenario to meet their criteria for 'sustainable development'.

## The Greenpeace 'fossil-free energy scenario'

Perhaps the most optimistic scenario describing the prospects for renewables by the end of the 21st century is the 'fossil-free energy scenario' (FFES) developed for Greenpeace International by the Stockholm Environment Institute (Lazarus *et al.*, 1993). The Greenpeace scenario's assumptions on population increase and economic growth are similar to those of the RIGES and WEC scenarios. The authors have reservations about these assumptions on environmental and social grounds, but have retained them in order to allow their scenario to be compared with others.

In the Greenpeace scenario, world primary energy demand increases to around 1,000 EJ by the end of the 21st century (Figure 11.8). All fossil and nuclear fuels are entirely phased out by the end of the

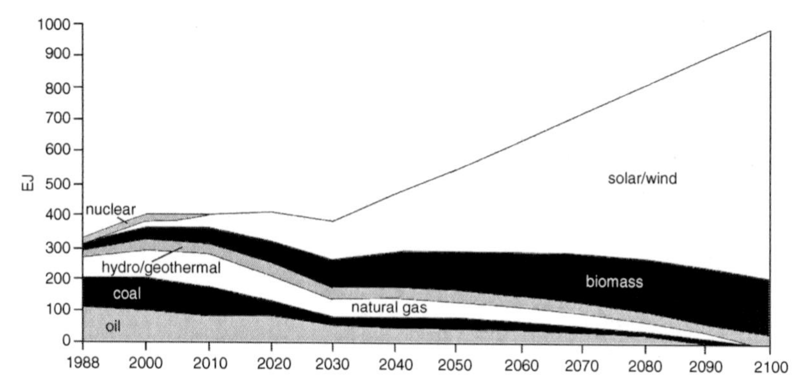


*Figure 11.8* The Greenpeace Fossil-Free Energy scenario. All fossil and nuclear fuels are phased out and replaced by a mixture of solar, wind, biomass, hydro and geothermal sources by 2100

*Source*: Lazarus *et al*., 1993.

period, and replaced by a mixture of solar, wind, biomass, hydro and geothermal energy. Improvements in energy efficiency, achieved using 'market or near-market' technologies, result in overall primary energy demand levelling out and then falling slightly around 2030, before rising again towards the end of the century.

The renewable energy technologies in the FFES supply mix by the end of the next century include: co-generation of electricity and heat (CHP) from biomass wastes; fuel cells for electricity (and heat) production; and increasing use of wind turbines, photovoltaics and solar thermal electric power generation. Hydrogen, produced by electrolysis from solar/wind sources or from biomass, provides an increasing proportion of transport fuel and a means of storing power from intermittent sources.

## DREAM-World

As the discussion above has shown, the levels of world primary energy demand and associated atmospheric carbon emissions during the 21st century will depend on the magnitudes of, and the interactions between, a large number of very complex factors. In an attempt to simplify and clarify the main variables, the author has developed a simple computer-based model, DREAM-World (Boyle, 2000) in which energy related carbon emissions are expressed as the product of four key parameters:

- World population;
- Economic activity (per capita GDP);
- Energy intensity (primary energy per unit of GDP); and
- Carbon intensity (carbon emissions per unit of primary energy).

In DREAM-World, logistic equations are created by specifying the rates of increase (or decrease) in each of these parameters at their 'inflection points', and the dates at which these 'inflection points' occur. The model provides a highly-simplified, but reasonably accurate, simulation of the historic variations, and the likely future patterns of variation, in the four key variables. It also allows users to generate their own scenarios and compare them with three generic scenarios produced by the World Energy Council, Shell and Greenpeace (see above). The model is available on the world-wide web (see References below).

## Conclusions

In the scenarios described above, the main drivers of energy *demand* are, as implied above: population; economic growth (GDP/capita); and the energy intensity of GDP.

Population projections differ significantly between scenarios, with the earlier United Nations projections envisaging nearly 12 billion people in the world by 2100, but more recent projections suggesting a peak of just under 9 billion in mid-century.

Rates of growth in economic activity (per capita GDP) are fairly similar between scenarios, with most suggesting annual growth rates of 2–3%. Most of the scenario authors conclude that economic growth is beneficial because it brings greater affluence (and this in turn usually leads to lower birth rates in developing countries) and accelerates the rate of technological change, speeding the deployment of low and zero carbon technologies.

However, some environmental economists disagree (for example Douthwaite, 1992; Daly and Cobb, 1990), arguing that economic growth is detrimental to societies, bringing many hidden costs that outweigh its benefits. This controversy remains unresolved.

Energy intensity (primary energy per unit of GDP) is assumed in most scenarios to reduce steadily during coming decades at rates between about 1% and 3% per annum. This assumes that efforts to conserve energy and use it more efficiently prove successful to a greater or lesser extent, depending on the scenario. Assumptions differ between regarding the extent to which the tendency towards 'dematerialisation' that has been experienced in many developed economies will continue to reduce the energy intensity of production of goods and services; and the extent to which the 'rebound effect' reduces the effectiveness of measures aimed at improving energy efficiency. (For further discussion of 'dematerialisation' and of the 'rebound effect', see Herring in chapter 7 of this volume and Boyle, 2003)

Regarding energy *supply*, most scenarios envisage strong deployment of low and zero carbon (LZC) energy technologies in order to mitigate climate-destabilising carbon emissions. The rates of deployment vary between scenarios and depend on assumptions about rates of improvement in the economic competitiveness of such technologies, and the policy measures that might be introduced to encourage their development and dissemination.

The key LZC technologies are: renewables; nuclear power; and 'clean' fossil fuel technologies incorporating carbon capture and storage.

Many of the renewable energy technologies, such as wind, photovoltaics, solar thermal, hydro, biomass, and geothermal power have been widely demonstrated and many (such as hydro) are fully mature, though some are only now entering mass production and deployment. A few are still at an early stage (wave & tidal stream) and need additional research and development. Most of the 'new' renewables need further mass deployment to further reduce costs.

Nuclear fission energy technologies are proven and widely deployed, but persistent concerns about cost, waste disposal, safety, nuclear fuel supplies, proliferation and terrorism risks remain. Improved nuclear reactor designs may help to mitigate some of these problems but are not yet fully demonstrated. Nuclear fusion is still many decades away from demonstrating a commercial scale prototype. (For further discussion, see Elliott, 2007).

'Clean' fossil fuel combustion, incorporating carbon capture and storage (CCS), although partially demonstrated on a relatively small scale (as in the Norwegian Sleipner project) needs to be more widely deployed at full scale, and costs need to be substantially reduced. Even when this has been achieved, some problems are likely to remain. With current technologies, only about 85% of $CO_2$ emissions can be economically captured; and in the case of CCS from coal, there remains the persistent problem of mining deaths and accidents, in the case of deep mines, or environmental despoliation, in the case of surface mining.

In the scenarios surveyed above, those that lead to the lowest atmospheric carbon emissions during the 21st century are generally those with the highest proportions of energy from renewable sources, coupled with rapid improvements in the efficiency of energy use. Renewables and energy efficiency ought therefore to be given the highest priority in national and international programmes of research, development, demonstration, deployment and dissemination.

There seems little room for doubt that a rapid deployment of renewable energy and energy efficiency improvements can – and should – play the leading role in enabling the world to make the vital transition to a zero-carbon energy future. Doubts remain, however, about the willingness and ability of national and international political and economic systems to implement the measures required to make such a transition a reality.

## References

Bos, E. and Vu, M.T. (1994) *World Population Projections: Estimates and Projections with Related Demographic Statistics*, 1994–5 edition, World Bank, Washington, DC.

Boyle, G. (2000) 'DREAM-World: A Simple Model of Energy-Related Carbon Emissions in the 20th and 21st Centuries', *Energy and Environment,* Vol. 11, No. 5, pp. 573–85. The DREAM-World model is available *via* the Open University Energy and Environment Research Unit web site, at: http://technology.open.ac.uk/eeru/index.htm

Boyle, G. (2004) *Renewable Energy: Power for a Sustainable Future,* Oxford University Press, 496pp.

Boyle, G. (2003) 'Introductory Overview' in Boyle, G., Everett, R. and Ramage, J. (eds) (2003) *Energy Systems and Sustainability,* Oxford University Press, pp. 1–54.

Boyle, G. (2007) 'A Tale of Two Countries: Non-Nuclear Sustainable Energy Futures for the Germany and the UK', in Elliott, D. (ed.) *Nuclear or Not?* Palgrave Macmillan, pp. 183–96.

Douthwaite, R. (1992) *The Growth Illusion,* Green Books, p. 367.

Daly, H. and Cobb, J.R. (1990) *For the Common Good,* Green Print, p. 482.

Elliott, D. (ed.) (2007) *Nuclear or Not?* Palgrave Macmillan.

Federal Ministry for the Environment, Nature Conservation and Nuclear Safety (BMU) (2004) *Ecologically Optimised Extension of Renewable Energy Utilisation in Germany – Summary,* BMU, Berlin, 48pp.

Goldemberg, J. (ed.) *et al.* (2000) *World Energy Assessment,* United Nations, New York (see also update supplement published in 2004).

Herkstroter, C. (1997) *Contributing to a Sustainable Future – the Royal Dutch Shell Group in the Global Economy,* paper presented at Erasmus University, Rotterdam, March 17th 1997, Shell International, London, 9pp.

International Energy Agency (2006) *World Energy Outlook,* IEA, Paris.

International Energy Agency (2003) *Energy to 2050: Scenarios for a Sustainable Future,* IEA, Paris, 224pp.

Intergovernmental Panel on Climate Change (IPCC) (2000) *Special Report on Emissions Scenarios* (SRES), available from http://www.grida.no/climate/ipcc/emission/index.htm.

Johansson, T.B., Kelly, H., Reddy, A. and Williams, R. (eds) (1993) *Renewable Energy: Sources for Fuels and Electricity*, Island Press, Washington DC, p. 1160.

Lazarus, M. *et al.* (1993) *Towards a Fossil Free Energy Future: The Next Energy Transition*. Stockholm Environment Institute, Boston Center.

Lutz, W., Sanderson, W. and Scherbov, S. (1997) 'Doubling of World Population Unlikely', *Nature*, 387 (6635), 803–5.

Nakicenovic, N. *et al.* (eds) (1998) *Global Energy Perspectives*, Cambridge University Press.

Royal Commission on Environmental Pollution (RCEP) (2000) *Energy – The Changing Climate*, The Stationery Office, London, p. 292.

Shell International (1995) *Evolution of the World's Energy System, 1850–2060*, Shell International, London.

Shell International (2001) *Energy Needs, Choices and Possibilities: Scenarios to 2050*. Shell International, London, p. 60.

Schrattenholzer, L., Mketa, A., Riahi, K., Roehrl, R. *et al.* (2005) *Achieving a Sustainable Global Energy System*, ESRI Study Series on the Environment, Edward Elgar, Cheltenham, UK, 209 pp.

Schellnhuber, H. J. (ed.) (2006) *Avoiding Dangerous Climate Change*, Cambridge University Press. PDF version downloadable from: http://www.defra.gov.uk/environment/climatechange/internat/dangerous-cc.htm

# 12
# Sustainable Lifestyles of the Future

*Peter Harper*

In the great debates over climate change there is always a ghost at the banquet – or is it an elephant in the room? It is the 'lifestyle' question: will we be obliged to change our aspirations, habits and patterns of life in fundamental ways? Or to put it less menacingly, would it be best if we did so?[1] It hardly needs to be said that serious suggestions of lower (or even limited growth in) material standards go against the grain of the modern consumer project, and are unwelcome to governments everywhere, not to mention their electorates. It is unsurprising then, that the vast bulk of policy, investment and research goes into 'technical' means of reducing carbon intensity that leave our customary ways of life, and dreams for the future, substantially unchanged: *vide*, most of this book.

The UK policy context is well known: the government, prompted by a Royal Commission report (RCEP, 2000), has set a target of 60% reduction for UK greenhouse gas emissions by 2050, relative to 1990 (HM Government White Paper, 2003). In the discussion below I shall refer to this as the 'UK 2050' target. This is based on the assumption that 550 ppm of carbon-dioxide equivalent ($CO_2e$) greenhouse gases (GHG) is a 'safe' atmospheric level. As of this writing, opinion seems to be moving towards the view that 550 ppm is too high, largely on account of potential feed-back effects, and that strenuous efforts should be made to stabilise at some lower level. A consensus is palpably emerging that UK emission reductions of 80% or even 90% will actually be required by 2050 (e.g., Tyndall Centre, 2005; DEFRA, 2004; Eurometrex, 2005; Hillman, 2004). The Stern Review, marking an historic and authoritative *volte-face* within the community of orthodox economists, also acknowledges that the range of possible outcomes embraces far more radical targets than UK 2050, declaring flatly that 'global emissions will have to be between 25% and 75% lower than current levels by 2050' (Stern, 2006).[2] In the discussion that follows we shall call the round figure of 80% reduction for the UK the 'Consensus 2050' target. Although this target is not specifically endorsed by the UN-sponsored Intergovernmental Panel on Climate Change, the IPCC's most recent data strongly imply that it is closer to the likely 'real' requirement than UK 2050 (IPCC, 2007). Achieving this stricter level would require policies that go far beyond Business As Usual.

It should be acknowledged that Business As Usual does deliver *something*. 'Energy intensity' – level of emissions per unit of economic output – does tend to go down 'autonomously' in post-modern economies (Kaufman, 2004; Harper and Todd, 2004). Yet in spite of steady reductions in both energy- and carbon-intensity (Boyle *et al.*, 2003), the crucial indictor – actual carbon emissions – declines only slowly if at all (DEFRA, 2006), and remains stubbornly resistant to the nosedives prescribed by 'consensus' climate scientists. The 'scale factors', broadly summarised by the Gross Domestic Product, are simply growing too fast.[3] Unless intensity-reduction starts to perform much better than it has in the last 20 years, we are mathematically obliged to pay more attention to the matter of aspirations and lifestyles. Along with the distributional question of *who* gets to consume how much, both within British society and across the world, this seems to be a bullet we will have to bite sooner or later (Meyer and Hildyard, 1997). Why not sooner?

Who should take the initiative here? The ordinary citizen tends to look to government and the business sector to institute reductions in carbon intensity, but as we have already noted, their performance to date does not match requirements. When it comes to cultural and lifestyle changes the prospects are even worse. Neither business nor government would easily accept that it falls within their remit to mandate lifestyle changes for individual households. Further, their scope for action is severely constrained by institutional inertia, complex stakeholder pressures, risks to competitiveness, off-shoring, and fear of electoral backlash.

Who's left to pick up the baton? Inevitably the spotlight must shift to *households* (Noorman and Uiterkamp, 1998). In a democratic market-based society, ultimately everything – production, government functions, services – is done, however indirectly, for the benefit of consumers, most of whose consumption takes place in the framework of a family unit, or more generally, a *household*. So as a matter of principle most of the consequences of consumption are ultimately the responsibility of households. Naturally most householders do not see it that way! But *some do*, and I would like to focus particularly on the subset of innovating households that do accept a large measure of responsibility for 'their' carbon emissions, and which seek creative ways to reduce them by unilateral actions within their own sphere of influence.

What share of emissions can usefully be attributed to households? The answer I would like to offer is *sources where households have a critical choice in the matter that cannot be easily made in other sectors*. Consider the following breakdown (Table 12.1) of greenhouse gas emissions attributed to different parts of the economy by the Office of National Statistics, including imports (Francis, 2004).[4]

About 35% of the total is *directly* emitted by householders for household energy and transport. Clearly when you boil an egg on the gas-stove energy is being used, and $CO_2$ emitted right under your nose. It is only a mite less direct when the gas central heating fires up or you start the car. With electricity the actual emissions are obviously remote, but can be related accurately to actual usage, as we know from our fuel bills. To what degree can we say these emissions are the 'responsibility of' householders?

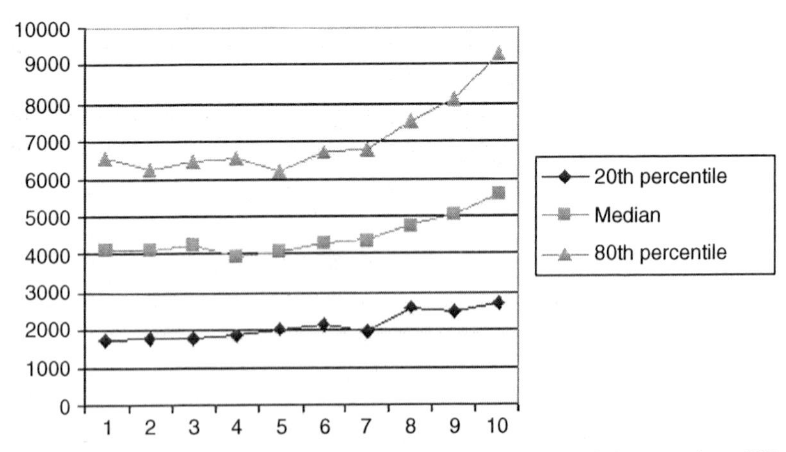


*Figure 12.1* Annual carbon dioxide equivalent emissions in kilograms from UK households, ranked according to income

*Note*: Income groups are ranked in 10% bands along the horizontal axis ('deciles'). The vertical axis shows direct household energy-related $CO_2$ emissions per year. Households in each income decile are ranked according to their domestic carbon emissions in 1% bands ('percentiles') and the cases selected are the 20th, 50th ('median') and 80th bands, indicating the emissions range from the middle 60% of the UK population.

Figure 12.1, based on dated statistics[5] but probably robust enough (Dresner and Ekins, 2004), shows house-energy carbon emissions as a function of income decile for households of equivalent size. This particular category of emissions varies rather little across income classes, but varies enormously *within* each class. To a certain extent this reflects life-cycle stages and household composition, but these factors cannot account for such a large ratio (>3) between the 20th and 80th emission percentiles (the difference between the 10th and 90th percentiles would be even greater). There is a strong suggestion here that 'habitual low-carbon patterns of domestic life' can reduce emissions by 50% or more without any special effort or specific measures. Imagine then what can be done with greater awareness, judicious behavioural shifts, and appropriate technical support.

In contrast, personal transport is much better-correlated with income, with the top quintile of earners travelling about three times further than the bottom quintile (Department of Transport, 2005; Korbetis *et al.*, 2006). A great deal of contemporary travelling is a 'secondary good', inessential for normal life but something you do with extra disposable income, including long commutes. In both these cases, then – house energy and transport energy – the level of emissions is largely a matter of personal choice, and can be properly regarded as the responsibility of householders. The business sector, however, should take principal responsibility for freight carriage and for trips carried out on behalf of an employer.

Now consider the other items in Table 12.1. Carbon emissions associated with food, goods, capital stock, investment and services, are *indirect*; that is, they are remote from the actual user. Clearly energy is used, and emissions generated, but it is harder to attribute a given level of emissions to a particular product or service, or particular actors.

***Table 12.1*** **Greenhouse gas emissions from various sources in tonnes of carbon dioxide equivalent**

| | All UK, Mt per year | Per Household, t per year | Per Head, t per year |
|---|---|---|---|
| Household energy | 100 | 4 | 1.7 |
| Personal mobility | 150 | 6 | 2.5 |
| Food, goods and services | 340 | 13.6 | 5.6 |
| Investment and infrastructure | 130 | 5.2 | 2.2 |
| TOTAL | 720 | 28.8 | 12 |

Part of this – direct government functions and the major infrastructure of the economy, schools, hospitals, roads, armed forces, public services etc – is completely beyond the direct control of householders or indeed the business sector. This part of the total emissions should then properly be attributed to the government. What about the private sector itself? Although businesses like to pretend they simply respond to consumer demand, they too have choices that are hidden from, and uncontrollable by, consumers, especially regarding the manner in which goods are created and delivered. They must therefore accept responsibility for a certain proportion of indirect emissions. But not all: most of the food, goods and services that householders buy are not strictly essential. As with transport, a large proportion are 'secondary goods', inessential cultural items the purchase of which increases rapidly with income. If these are choices made by householders, they must take responsibility for them.

Having looked at many attempts to track the pattern of carbon emissions through the UK economy,[6] it seems to me that the 'responsibilities', judged by the loci of choice, are disposed approximately as follows:

Direct use (basically consumers' responsibility, with a minor business element) 30%
Indirect (basically consumers' expenditure and purchasing choices) 30%
Indirect (responsibility of the business sector) 20%
Infrastructure (responsibility of government) 20%

Of course the government always holds the ring, or should. In principle it can influence choices in any of these spheres through a vast range of legal, financial and ideological instruments. It should participate actively in appropriate international agreements. It should use its unique access to information to look well ahead, engage the public in honest debate and seek benignly to carrot-and-stick everybody in the right direction.

The business sector too, could in principle make a major difference by, on the one hand cleaning up its own operational acts, and on the other providing appropriate products and services that allow consumers to reduce their emissions (Hawken, Lovins and Lovins, 1999). For example, the dark arts of 'persuading people to want something they don't really need' could be re-directed to the promotion of low-carbon products (Cyberium, 2006). Yet it would be rash to expect too much in the way of unilateral contributions from the business sector: this is simply not its job. It is the motor of the economy, not the steering wheel.

We should not neglect other influential 'estates': the media, the universities and the institutions of civil society. Taken together, these and the government and business sectors could be considered the 'Public Sphere'. Within all parts of the public sphere there are dynamic, creative and courageous elements striving to link up and move forward. They all have essential roles to play, and can all influence household consumption.

But in the end most of the 'loci of choice' lie with households. True, most householders will find this uncomfortable and habitually look to the public sphere to take most of the strain; but here I want to focus on households that have, as it were, lost faith in the ability of the Big Players even to fulfil their own proper role, much less bring about a full-fledged top-down technical solution. Such households are ready to roll up their sleeves and get on with it, *irrespective of what government or industry might do*. They often ask this kind of question: 'If all 3 billion households [of the mid-21st century] lived like this, would it stabilise emissions below the critical level'? Or to put it another way, to each prospective new technology or pattern of life they apply a simple but resonant touchstone: 'What if *everyone* did it?'.

For the rest of this chapter, therefore, I shall focus on what individual householders can accomplish, either by household-scale low-carbon technologies, community-scale technologies, or by the adoption of low-carbon behaviours.

If householders take full responsibility for per capita national emissions onto their own shoulders, what is their 'sustainable fair share'? For the UK 2050 standard is it on average around 4.5 t carbon dioxide equivalent ($CO_2e$) per head per year, and 2.3 t for the Consensus 2050 standard (Francis, 2004; Hillman, 2004).

I would like to explore approaches to these 'sustainable fair shares' through some speculative scenarios focusing on the year 2030. I choose this date because it is not really so far away – closer in fact than the shadowy era when many of the contributors to this book first started thinking about these matters – and we can make plausible guesses about the likely technical and political possibilities. At the same time it is a useful stepping-stone to the signal year 2050 which is the focus of our chosen target levels.

Imagine two hypothetical British households of 2030, each striving to get below the target levels. For the sake of representativeness, both the hypothesised households have two adults and two teenage children. The implicit sharing of resources in such households lowers their emissions somewhat relative to the average household of 2.4 people, so their household targets are (in the roundest of round figures) 16 t per year for the UK 2050 standard, and 8 t for the Consensus standard.[7] How do they perform in terms of carbon emissions? Let's assume they can both benefit somewhat from expected improvements in technology and infrastructure.

The first household is generously resourced and its members are able to buy virtually any low-carbon goods or systems applicable at the household scale. Meanwhile *they insist on maintaining customary habits and living standards commensurate with their income*. I shall call them Well-Off Techie environmentalists or **WOTs**. The other imaginary household has modest resources and chooses an almost entirely cultural route to sustainability. They cannot buy sophisticated emission-reducing facilities or equipment, but *are* willing to make major changes in their habits and aspirations. I shall call them Low-Income Lifestyle environmentalists or **LILs**.

## The WOT household

It has access to capital, and its income is in the ninth decile, around £60,000 a year (Institute for Fiscal Studies, 2006). It has a custom-made 'eco-house', super-insulated with advanced glazing and smart heating-controls. It is generously-sized. 'Normal' comfort standards are maintained throughout the year. Appliances are of optimum efficiency and also controlled by smart systems including load management.

The household has two vehicles, a lightweight 'hypercar' of the kind proposed by Amory Lovins of the Rocky Mountain Institute (RMI, 2006) running on a mixture of fossil and biofuels, clocking up 20,000 km a year, and an electric car using Green Tariff electricity doing 5,000 km a year. Other trips are by bicycle, and public transport at 3,000 km per person per year. The house is stocked with high-quality, long-lasting furniture and other goods using as far as possible sustainable-source materials. On-line information and 'electronic paper' really have (at last) replaced a great deal of printed matter. Food is virtually all organic, but conventionally sourced, principally from supermarkets, although much of their food is ordered on-line and delivered. They have a private garden of 200 m2, used for recreation. Three holidays by air are taken each year, two of 2,000 km round trip with two travellers, the other *en famille* of 10,000 km with four. Travel for business is not included.

How far have they got towards sustainability, in comparison with today's average?

Some things evidently *can* be 'fixed' by household-scale technology. The *Passivhaus* standard (Passivhaus, 2005) in Germany, or the AECB's Silver standard (AECB, 2006) in the UK are clearly achievable, if expensive. That is however what we expect of wealthy environmentalists: that they can reduce their impacts by sheer main force of selective spending. In the hypothetical case considered here, they have indeed achieved a massive reduction in house energy. It could easily be reduced to under 2 t (Heating 400 kg, water heating 600 kg, cooking lights and appliances 800 kg).

A 'hypercar' using conventional fuels achieves twice the conventional mileage, but this is after all a well-heeled household with two cars, and the expected mileage would be quite high. If the hypercar achieves GHG emissions of 80 g/km – 'normal' vehicles rate between 120–200 g/km – this gives a total of 1,600 kg per year on conventional fuels. Amory Lovins envisages the hypercar eventually running on hydrogen. Probably by 2030 wealthy households will be able to purchase biofuels or even renewably-generated hydrogen on the open market. Let us suppose they are able to meet 50% of their fuel demand from renewable sources, giving 800 kg for the hybrid. By 2030 Green Tariff electricity might really mean something, so 40 g/km would be a reasonable guess for the electric vehicle, giving 200 kg a year. The total would be 1,000 kg for cars. Then there is public transport. Wealthy families tend to travel a fair amount by train. Users cannot unilaterally alter the impact per km travelled, although probably by 2030 there would be a substantial renewable element for public transport. The expected emissions would be about 400 kg, giving a total of 1,400 kg for surface transport, relative to an expected level of 2.8 for a four-person household today, giving a total of 1.4 t. In spite of higher mileage, we can see how, in principle at least, improved technology can reduce emissions.

Air travel, however, is a much harder nut to crack: This is a category particularly reflective of household income, but we have assumed the household takes only three plane trips of middling distance. In this area *the prospect of technical fixes appears severely limited*, even for the wealthiest (Upham *et al.*, 2003). Even the much-vaunted hydrogen, used as an aircraft fuel, might not help much, injecting water vapour into the upper atmosphere just as conventional fuels do now and creating climate-forcing effects (Sausen *et al.*, 2005). Let us suppose however that efficiencies have improved, and emissions per kilometre are somewhat lower by 2030. The shorter two-person trips would create about 0.6 t $CO_2e$ each, the longer four-person trip 5.2 t, a total of 6.4 for the year. It is interesting that taking family holidays by air (or indeed other forms of public transport) the emissions of all persons are added, whereas with a car they are shared. We shall not count work-related trips, which fall into another part of our carbon statistics.

The emissions attributable to household goods, both durable and consumable, are difficult to calculate. According to the Office of National Statistics (Francis, 2004) household goods account for about 1.3 t $CO_2e$ per head, and leisure goods and services around 1.6 t. For a four-person household, much of the first category is shared, but less of the second. I shall assume the baseline household goods emissions as 2.6 and the leisure goods and services as 4.8, giving 7.4 t for the whole family. What does the wealthy household do if it wants to unilaterally buy low carbon goods and services? In general household emissions are positively correlated with income, so we would expect the default value to be considerably higher than the present average. The WOTs are unlikely to spend *less* – virtually all households spend what they get – but they would spend artfully, on durable products made from sustainable materials. Nothing but the best! They would probably substitute many services for goods, and take advantage of Product Service Systems (UNEP, 2003). They would doubtless dutifully separate waste for recycling, but probably not re-use or repair very much. Let us assume that the WOTs two-fold greater-than-average income commands an extra 50% of goods and services (Sutherland, Taylor and Gomulka, 2001) to 11 t, but that discriminating choices could more than halve the expected emissions, to 5 t, along the lines suggested by Hawken, Lovins and Lovins (1999). This would be aided by a kind of reverse-rebound effect, since the extra costs of 'buying green' would reduce income available for non-green choices (Hertwich, 2003a).

Food is another problem area for technical fixes: by simple purchasing choices they could eat mostly organic food, and this would somewhat reduce carbon emissions through reduced fertiliser use and $N_2O$ emissions from soil (Pretty, 2005). However with a conventional diet they would still indirectly emit from food production, particularly from animal products, processing and transport. Composting organic wastes and growing some vegetables in their own garden would make only a small dent in the total because fresh fruit and vegetables are already low-carbon items. The 'default' level for four people would be 11.6 t. It is hard to see how they could get below 9 t from food, without a radical change of diet, basically eating much less meat, fish and dairy products, which for the moment our assumptions for the WOTs rule out.

Their total emissions are therefore just under 24 t per year. To this we have to add the fixed element that represents investment in the UK economy and its infrastructure, that we all have access to, and is beyond any direct influence by consumers. We have calculated this at a minimum to be 2.2 t per head, or 8.8 for a contemporary household of four. By 2030 we might expect this to have been reduced by general decarbonisation policies to perhaps 1.5 t per head or 6 t for the household. This gives a grand total of 30 t. To get down anywhere close to the UK 2050 target level of 16, and *a fortiori* the 'real' target of 8 clearly requires substantial changes in the general provision of goods and services. In spite of excellent performance in some important aspects of their lives, the WOTs cannot adequately decarbonise themselves by sheer consumer power or technological main force. If they want to be truly sustainable they must rely on the 'public sphere' to decarbonise all their inputs, or else change their habits.

## The *LIL* household

Now let us turn to the other end of the spectrum, where cultural and behavioural measures are deployed to achieve a sustainable level of emissions. The Low Income Lifestyle house is smaller, with a modestly-sized garden. Suppose they also have 2 + 2 inhabitants but an income in the second decile (equivalent to about £15,000 at 2005 prices). Their project, as it were, is to try and get under the target of 8 t without the benefit of any special technology, living a modestly comfortable modern life. They may use anything that would be standard and affordable in 2030, but otherwise they have to adapt as best they can.

With respect to household energy they would be hard pressed to match the performance of the WOTs without any technological help. We might suppose however that by 2030 nearly all existing dwellings have been brought up to the kind of insulation and airtightness standards mandatory for new-build today. The German Federal Government has already committed itself to a 20-year programme to do just this, and in all likelihood the UK will get round to it sooner or later – probably later, but soon enough for our scenario. As a result we could postulate space heating emissions of around 1.5 t based on careful use, warm clothes and low temperatures. The 'responsible' environmental view is that we should aim to run homes at no lower than 19°C (Boardman *et al.*, 2005), but from the LIL perspective this is a good opportunity missed. My own investigations (Harper, 1999) suggest that lowered temperatures down to 17°C or even 16°C are a fairly quick and reliable behavioural route to lowering household emissions if the inhabitants can adapt, and they usually can. The logic here is very simple: just keep the temperature as low as you can handle without undue discomfort. Hot water could be reduced from 2.1 to 1.5 t, and cooking and appliances to 2 t, simply by minimising their number and using them carefully. By 2030 it is likely that all available appliances will be at the top end of the current energy ratings scale. So the sum total for household energy would be 5 t, or 1.25 t per head.

Going on to personal transport, here the LILs adopt the standard car-free package of making all but a very few journeys by foot, bicycle and public transport. Perhaps they occasionally hire cars for awkward journeys, or more likely would be members of a car-share club (Carplus, 2006). They take a few long-distance trips by public transport each year, but essentially they live locally and don't travel much. Allowing for partial decarbonisation of public transport by 2030, their emissions could be 400 kg, or 100 kg per head. Finally on air travel, as a matter of principle, this household never flies. Holidays are taken closer to home, and by surface transport.

Their diet, almost fully under their control, is carefully calculated to minimise emissions. It is vegan, organic, with a high proportion of local foods and a minimum degree of processing. They eat a high proportion of dried goods, and raw food. They buy in bulk, share orders with friends and neighbours, and take great care not to incur unnecessary carbon emissions through collection and deliveries The result is emissions from food at around 3 t per year, a really big difference from the reference household's 11.6 t, highlighting in dramatic terms the large (and largely unnecessary) contribution of the food system to current national emissions.

When it comes to household and leisure goods and services, according to statistics derived from the Office of National Statistics, the figure would be 11–12 t per year for a typical household of this size. However in this case emissions would be limited simply by low expenditure, and of course this particular household would be making conscious choices in nearly every sphere, and we estimate a highly-aware household could cut this to 3.

The total is therefore 5 + 0.4 + 3 + 3 = 11.4, which looks very promising. However we have to add the estimated infrastructure term of 6 t, giving an overall total of 17.4 t. Perhaps it is a little harsh to suppose that all households 'deserve' an equal share of the UK's infrastructural background and therefore must take equal responsibility for its carbon emissions, but this must surely be the default approach. As a proportion of the total, it bears heavily on low-emission households, but they are still using the roads, hospitals, offices, NGOs, schools and government functions, perhaps more in some ways than the WOTs who might well favour private sector provision.

The results of this admittedly speculative investigation are summarised in Figure 12.2. Both hypothetical 'green' households do better than today's average, and the LIL household of 2030 almost meets the national target for 2050. It is not that we are expecting national targets to be met entirely through household activity; but this result means that *if* all UK households performed at this level the government would meet 'its' own target with minimum effort. Even the LIL household however, does not single-handedly reach the Consensus 2050 target of 8 t, arguably the 'real' touchstone for environmental sustainability, and this raises the question, what next? Are these the practical limits for such households? Could they do even better? Where might they be in 2050?

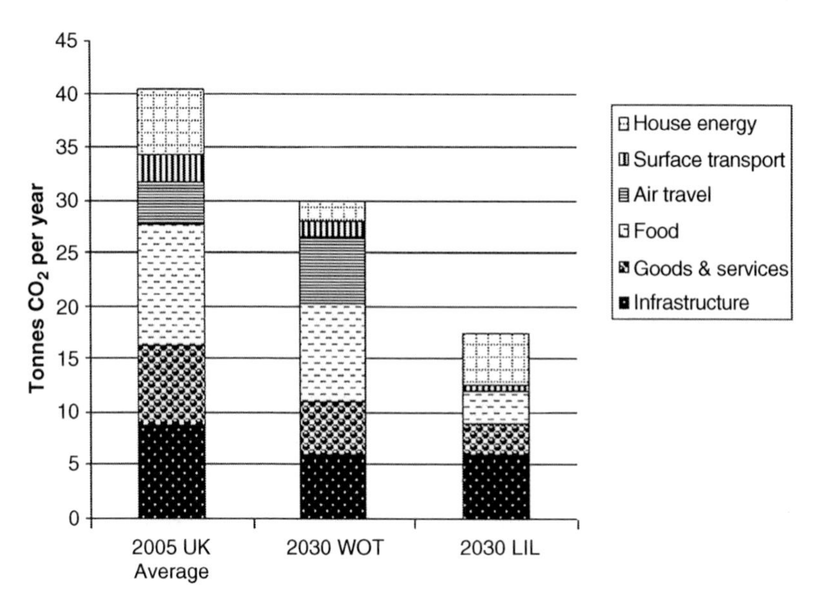


*Figure 12.2* Emissions for hypothetical 4-person households in tonnes of carbon dioxide equivalent per annum

*Note*: For context, the UK 2050 target is around 16t for households of this size.

The results could be 'improved' if we were to deduct a proportion for which the corporate sector should take responsibility. But the point of this particular exercise is to highlight households that are attempting to go it alone, thumbing their noses at both government and business. It is also a way of gauging how much the large-scale 'public sphere' might be required to contribute if households do nothing.

It is difficult to see how the LILs could go further in the direction of sensible living and sturdy frugality without disappearing off the radar of modern life. What they need most is better domestic technology for house energy. It is agreeable to speculate that as awareness deepens regarding the climate dilemma, LIL-type households will be recognised and supported as 'environmental champions' by government agencies. Their houses might be comprehensively retrofitted with high levels of insulation, airtightness, solar water heating where appropriate, state of the art boilers and appliances, smart controls, perhaps micro-CHP. A well-conceived basket of measures, plus an improvement in the background infrastructural contribution, could bring the LIL household down to 11 or 12 t by 2050, well inside the official UK target. Further, even a low-income household, if motivated by a desire to reduce personal emissions, might calculate that money is better spent on offsite technologies such as forestry, biomass crops or wind farms. One per cent of the income of the hypothesised LIL household, invested annually in a wind-farm such as that described in Box 12.1, would offset about 3 t a year, bringing the household close the Consensus threshold.

One of the most important avenues for LILs to explore is the possibility of systems that perform best at a neighbourhood or community level, for example bio-fuelled district-heating (Energy-Saving Trust, 2004), co-generation plants, composting, food-sourcing, even house-building. Such initiatives can emerge spontaneously in pre-existing communities, sometimes with impressive results, such as the Car-Share system described in Box 12.2. The question then arises, can we not create *new* settlements where the respective low-carbon systems are built-in at their own optimum scale? This is the province of the 'Ecovillage' concept (Dawson, 2005) with quantifiable carbon-reduction (Meltzer, 2005). These community-level possibilities have been effectively invisible to mainstream planning policy, and opportunities are being missed. Every new settlement that fails to achieve the strictest energy standards is another millstone round our necks that will have to be retrofitted or laboriously offset sooner or later.

*Box 12.1* **Bro Ddyfi Community Renewables**

The BDCR Ltd is a social enterprise set up to develop renewable energy installations in the Dyfi Valley in mid-Wales. A quick glance at its web site www.bdcr.org.uk is very instructive. Membership is open to anybody for a small fee, and members may buy shares in RE projects. The first venture was a 75 kW second-hand Danish wind-turbine with 59 shareholders, commissioned in 2004. Output is sold to the nearby Centre for Alternative Technology, who use what they can and sell on any surplus to the grid. This was the first community wind turbine in the UK fully managed by a local community group through planning, sourcing, purchase, site-works, installation, and commissioning.

Following this success, it was decided to replace a failed prototype machine for which planning permission had already been obtained, using another, 500 kW, second-hand turbine linked directly to the grid. This time there were over 200 investors, and the machine was installed in September 2006. The initial cost was around £250,000 for purchase and installation, and maintenance is expected to cost around £2,000 a year over a 20-year lifetime, giving a lifetime cost of around £300,000. At a capacity factor of almost 25% it is expected to generate about 1 GWh per year or 20 GWh during its lifetime. This can be set in various contexts. It could be said to provide the electricity requirements of the subscribers from a patently renewable source. Alternatively it could be regarded as an 'offset' by subscribers, since in the grid it would displace the average mix of about 75% fossil fuels, saving perhaps 0.3 kg $CO_2$ per kWh, or about 6,000 tonnes over its lifetime. To relate this to our hypothetical households, the LILs could invest 1% of their income (£150) each year, displace 3 tonnes and meet the IPCC target. In terms of disposable incomes the WOTs can probably afford 3% or £1,500 a year. If they were investors in the Bro Ddyfi machine it would buy them 30 'nega-tonnes' a year and get them handsomely under the bar. It could be argued that this level of targeted self-taxing is what wealthy but green-minded people should be doing with their money for ethical reasons. But of course there is also a financial return on the investment in terms of electricity sales and carbon credits, calculated to pay back in 15 years. These receipts can themselves be re-invested. So it does not strictly require an ethical motivation. Although in terms of national or indeed global needs such one-off projects might be considered insignificant, on a per head basis they show what relatively small sums can actually achieve.

### *Box 12.2* The Machynlleth Car Club (MaCC)

Car Share Clubs suggest a promising way to maintain customary levels of mobility with much lower carbon emissions. They can be commercial operations in which the vehicles are owned by a company, or a voluntary arrangement between householders to share a relatively small number of vehicles. Members usually pay per mile of use. Sometimes there is a joining fee. Payments to the club cover all fixed costs such as purchase of vehicles, tax, insurance, and also variable costs such as repairs and fuel. In addition, they usually cover the expenses of an administrator.

However high the level of car use, overall costs are lower because the fixed costs are shared. But in practice the economic advantages go much further because the 'perverse incentives' that normally result in unnecessary car use are reversed. Because conventional households pay all the fixed costs of a vehicle themselves, this is already a sunk cost, and the marginal cost of motoring is then just the cost of fuel. When this is compared with the costs of public transport it is often lower, so the cheaper option is to take the car. This is often given as a personal reason why public transport was not used for a particular journey ('much too expensive'). If however the full costs are reflected in the mileage rate the comparison usually goes the other way. Public transport is *actually* cheaper, but doesn't look like it. Take for example a single-person trip from Birmingham to London, a distance of 131 miles. In a vehicle doing 40 mpg, with petrol at £4 a gallon, the cost would be just over £13. A cheap-day single fare on the train would be £25. Although it usually takes longer by car, the difference in price might well make the train feel like an unwarranted luxury, especially as you have already paid the fixed costs and might as well take advantage of them. In contrast, the same journey at Car Club rates (35p a mile in the MaCC) would cost £39.30 – almost certainly enough to make you take the train. At first sight this might look like a disadvantage to the Car Club system, but this extra cost represents one that as a car owner you are paying anyway, but in a Car Club is shared. The difference is that the perverse incentives of car ownership are transformed into benign incentives. The observed result in the MaCC is that 80–90% of journeys are taken by other modes: walking, biking, bus or train. Only the 10–20% rump of really difficult journeys, or emergencies, or ones involving lots of luggage, are taken by car. The upshot is that car mileage is reduced to an average of around 1,000 per participating household, costing around £350 a year. Of course some members use the cars more, and all incur greater costs on the public systems than car-owners would. Still, it is much cheaper. Is it less convenient? Yes, in some ways: you have to make a booking on-line, and very occasionally all the cars are booked out. On the other hand, someone else takes care of all the worries about MOTs, insurance etc, and you have a choice of vehicles with different qualities depending on purpose.

In carbon terms this reduces emissions per household attributable to car use by about 80%, although we would expect some compensatory rise in public transport use. We should take into account also that many fewer cars are used. In terms of embodied energy this alone serves 10–20 tonnes CO2e per year.

A further step in Machynlleth is to use filtered used vegetable oil in a 50% mix with mineral diesel. This could in principle be increased to 100%, taking the members close to carbon-neutral motoring. Yes, obviously the supply of recycled vegetable oil is limited and this could not be generalised. Biodiesel would have to be produced by dedicated fuel crops, and this takes up land that competes with other uses, notably food. But at this mileage the requirement would only be about 0.2 ha per participating household. To put this in context, 25 million households in the UK would then require 5 million ha. This is admittedly a large area, but could occupy part of the 7 Mha currently used for grass and fallow, the 6 Mha of rough grazing and the 5 Mha used for growing animal feed that would be largely freed by the much lower livestock levels necessitated by a low-carbon food system. There are many disputatious details here, but it is enough to demonstrate that the members in Machynlleth at least, have come close to sustainable motoring without sacrificing the ultimate benefit of self-drive vehicles: to have access at short notice for urgent journeys, for the occasional trips that cannot be achieved in any other way (Sloman, 2006).

What about the WOT household? Could *they* do even better? Much as we might applaud the LIL household's achievements, we have to recognise that they have been accomplished at the cost of shaking some of the pillars of modern life. Certain standards are widely regarded as non-negotiable, and these include:

- 'Heedless comfort' in terms of indoor temperatures all year round.
- At least one private vehicle.
- A regularly high level of meat, fish and dairy products.
- At least one annual holiday abroad, and perhaps several.
- A steady flow of new consumer goods, both durable and consumable.

Any plan that outraged these expectations would be expected to suffer from at best, low take-up, or at worst, riots in the streets. This is where the WOT approach has its greatest value. The WOTs avoid challenging the 'non-negotiables'; they are recognised, respected even, as successful *bona fide* members of consumer society, and as trend-setting early-adopters of advanced technologies. We can assume that between 2030 and 2050 WOTs will continue to exploit the latest technical aids, but it is hard to see how they are going to reduce emissions associated with air travel, food, and consumer goods without some kind of cultural shift.

In each of these latter categories technical fixes are difficult, but for different reasons. In the case of food, the ONS and other sources suggest that the largest single impacts come from livestock (largely on account of enteric methane emissions), followed by food processing, fertiliser manufacture, release of greenhouse gases from soils, and transport (Food Climate Research Network, 2005). It is difficult to imagine purely technical solutions for all these, but if all food processing, transport and fertiliser production were performed with zero-carbon fuels it could reduce emissions from the food sector by 40–50%. Possibly the WOTs' most important contribution to sustainability could be the shift to an imaginative and sophisticated low-animal-content cuisine. This is not so difficult if done in stages. They would be excellent role-models. On a world scale sustainable diets are going to be politically essential (Reinders and Soret, 2003). If the WOTs adopted 'best practice' and ate meat or fish only on special occasions, or as a kind of condiment, they will have reduced their food emissions to about 4 t per year. Incidentally, this is the kind of diet recommended for optimum health by recent epidemiological studies in the US (Harvard School of Public Health, 2006; Harper, 2006).

Flying is even more intractable, since the rather slow improvements in efficiency per passenger-km are completely overwhelmed by the increase in the number of trips and distance travelled. Remember too, that according to the IPCC the temperature-forcing effect of flying is 2–3 times worse than the raw greenhouses gases emitted, on account of vapour effects in the upper atmosphere (IPCC, 2001, Sausen *et al.*, 2005; but see Forster, Shine and Stuber, 2006). There are only fixes of a very limited nature in prospect (Lee, 2003). WOTs have no alternative but to consider their aviation choices more carefully, or in the meantime to seek offsets, discussed below.

Moving on to consumer goods, these account for a large proportion of the indirect emissions. Whereas house energy and food are subject to saturation and are relatively income-inelastic (Dresner and Ekins, 2006), purchase of 'secondary' (i.e., inessential) goods is something that well-off households inevitably do. In carbon terms, money is dangerous stuff. Once the basic requirements are satisfied, wealthy households spend their incomes on items such as travel, larger houses, second homes and all manner of consumer and leisure goods and services. These can be very carbon intensive, and if money is not spent on one thing, it is spent on another, as Horace Herring points out elsewhere in this volume. In these spheres emissions are largely proportional to income (Vringer and Blok, 1995; Hertwich, 2003b). It is extremely hard for the wealthy to spend their way to ecological salvation simply through better products. One is tempted to postulate, in an update the original chroniclers might well have approved, that 'It is easier for a camel to pass through the eye of needle, than for a rich man to become sustainable'.[8] What are the poor WOTs to do?

Of course they should eschew pointless and profligate activities. But to ensure that the money saved is not simply diverted into equally carbon-intensive forms, they should use their surplus resources (extra buildings, for example) to propagate and show-case best-practice technologies, and help bring markets to maturity. They should invest in under-funded but promising fields. In the end their main hope is to be bailed out by a society-wide process of decarbonisation. They should therefore vigorously and materially support legal and financial changes such as environmental tax reform or carbon-rationing – and of course the international agreements without which such measures would have no meaning. We can assume that by 2050 the general background decarbonisation will have reduced the intensity of consumer goods substantially. In the meantime households *can* unilaterally invest in immediate carbon-displacing technologies to 'offset' their carbon excess. Offsets are a temporary measure and in the long run no substitute for genuine emission-reductions. But they are (surely?) better than nothing, and in the short term can be an effective way for the wealthy to launder their literally filthy lucre.

And they can do it on a grand scale. At a rate of about 20 kg per year per £ invested, it is obvious that wealthy people can fairly easily buy enough 'indulgences'.[9] Taking the case of our hypothetical WOT household, 1% of income invested each year in a scheme such as the Bro Ddyfi wind turbine (see Box 12.1) would displace 12 t per year. Supposing that by 2050 various household-level and background measures had brought the household emissions down to the official target

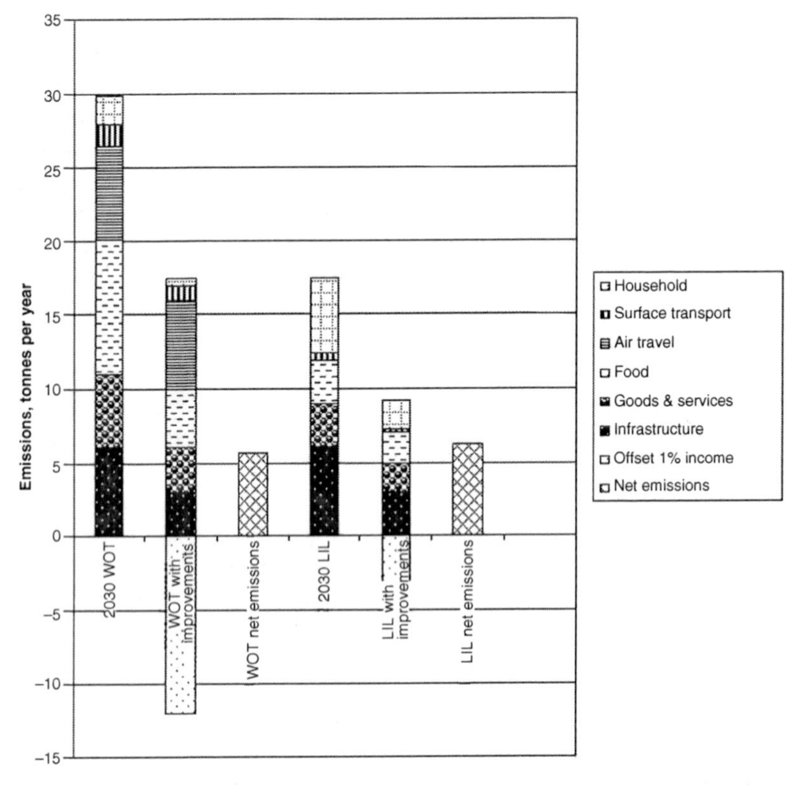


*Figure 12.3* Emissions from WOT and LIL households in tonnes of carbon dioxide equivalent per annum

*Note*: Here the basic and 'enhanced' annual emissions from the two hypothetical households are shown, with the effect of offsets purchased with 1% of gross household income.

of around 16 t, a 12-t offset would take them well below the Consensus target of 8. This is illustrated in Figure 12.3. In contrast, the LILs' 1% buys much less. But why should a wealthy household stop at 1%? If the LILs can do 1% the WOTs can easily do 3%. A 36-t offset takes them into net-negative emissions, and this should be a minimum standard for WOTs everywhere.

Note that the costs of these offsets are far higher than those customarily offered by on-line offset services undertaking to plant trees or invest in low-carbon programmes, particularly in developing countries, prompting the obvious question, 'if it's this cheap to offset our emissions, what is everybody worrying about?'. In principle there *are* good reasons to invest in such schemes in developing countries (Swingland, 2002), and they are indeed cheap (Kauppi and Sedjo, 2001). but there remain many procedural, ethical and legal question marks, and for the time being seriously low-carbon households generally prefer a guaranteed offset closer to home, ideally on a nearby hill. The important conclusion is that, in spite of this 'high cost' *it is still not economically crippling*. Real, tangible offsets derived from deliberate household investment in low-carbon technologies can be, in simple terms, a bargain. Scaled up, the same techno-fiscal realities underlie the relatively low costs of carbon-mitigation to a national economy envisaged by the Stern Review (HM Treasury, 2006). It can be done.

Such offset programmes amount to voluntary carbon taxes and are reminiscent of the medieval tithing system that supported the church institutions. Green-minded householders are usually willing to be tithed in this way, but in the long term a system of fiscally-neutral carbon taxes seems almost inevitable (Royal Society, 2002), perhaps with personal quotas to ensure an acceptable degree of equity (Hillman, 2004).

Returning to the LILs, they have obviously performed better than the WOTs in their own terms, and can claim to be close to the 'What if everybody did it?' standard of carbon emissions. They deserve credit and support. Nevertheless we can see there is a cultural problem of wider take-up. LILs need to devise smarter forms that will align better with the grain of modern consumer societies. Car-Share Clubs are case in point, and an example is described in Box 12.2.

The various trends and trajectories of the speculations, relative to the target levels are summarised in Figure 12.4. Without offsets, the WOT household, from a high start, reduces its direct and indirect emissions rapidly, and more or less reaches the government target levels by about 2040. The LIL household declines far more slowly, but from a much

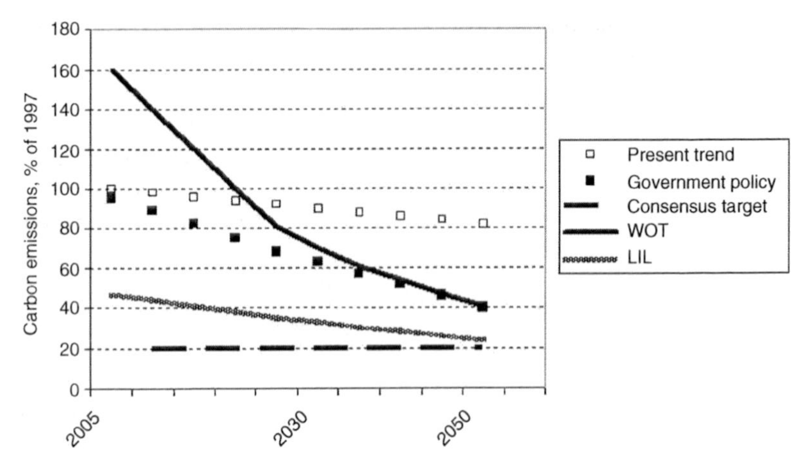


*Figure 12.4* Trajectories of various emission trends, UK 2000–2050

*Note*: Emission Trajectories. Open Square boxes show 'Business As Usual' trend in carbon emissions. Filled squares shows UK government intention for 60% reduction by 2050. Dashed line shows Consensus Target for UK emissions. Solid line shows hypothetical performance of WOT household. Hatched line shows hypothetical performance of LIL household. Voluntary offsets are not included.

lower start, and by 2050 has almost reached the Consensus target. I should emphasise once more that these trajectories have a largely rhetorical significance: it is not required that every individual household reach the targets on its own. Nevertheless the fact that a particular class of households appear to have done so, and that they could do even more with voluntary offsets, attests their potential importance to an overall strategy.

From some perspectives the LIL approach might be regarded as eccentric and of no more than academic interest. But unexpected support for it has appeared recently in the so-called 'Wellbeing Literature' (Layard, 2005; Jackson, 2005). This notes that the correlation between national income and reported levels of well-being is surprisingly weak. What then, these theorists ask, is economic growth *for*? Glibly put, all it seems to do is make the battle against rising carbon emissions harder, with little corresponding benefit except a vague feeling that surely things must be getting better (New Economics Foundation, 2005).

Of course the lifestyle issue is far more fraught, complex and paradoxical (Jackson, 2006). In some ways it is easy for a bohemian fringe to call for lifestyle change because it would cause them little difficulty.

They have carefully developed 'dematerialised tastes' and in many cases have spent years making appropriate adaptations. They often forget that for the vast bulk of the population nearly every aspect of life is intimately dependent on an abundant flow of low-cost material goods and services. But perhaps, at last, we are seeing the revival of a serious debate about the fundamental ends of a modern society; a debate that has recently belonged only to academics, the religious, and woolly-hatted idealists.

It seems to me that a positive and pro-active approach could turn many of these lifestyle questions from threats into opportunities, making an important contribution to a mixed techno-political strategy. It is not really a question of either-or. Synergies can be found between the two approaches. The top-down measures can make the bottom-up more efficient and palatable, while a dash of bottom up can sometimes be the secret ingredient that lubricates and catalyses changes in a centrally-mandated system.

I have tried to argue that if official policy does not effectively harness the potential of the household sector it will fail to achieve its targets. And let us remind ourselves that the stakes are high: they are *our* targets too. But we should not go to the opposite extreme, of relying completely on lifestyle changes. We cannot assume that the great majority of householders will be willing or able to match the performance of the cases described here Hence the pressing need for vigorous, thoroughgoing, government-led, society-wide measures of general decarbonisation. *Vide*, the rest of this book.

## Notes

1 The question has, of course, also provided the Green Movement with its principal fault-line.

2 If the world economy emitted 42 Gt CO2e in 2005, Stern's 25% reduction would mean 31.5 Gt in 2050. Divided by an expected population of about 8 billion gives about 4 t per head. A 75% reduction would give 10.5 Gt in 2050 or 1.3 t per head.

3 See for example an extended discussion in Chapter 7 of *The Stern Review*. (HM Treasury, 2006)

4 In comparison with most data-sets on carbon emissions, these figures might seem rather high, but they include an informed estimate of overseas emissions induced by UK consumption, and the extra forcing effects of aviation beyond the measured carbon emissions.

5 These data are adapted from the English House Condition Survey 1996. The survey is supposed to be conducted every five years, but data from the 2001 survey are still not available.

6 For a particularly detailed analysis of the actors influencing environmental impacts, see Lorek and Spangenberg (2001). See also 'Creating Low-Carbon Communities' www.bioregional.com
7 Allocating children the same share as adults might seem odd but in the present context is a necessary simplifying assumption. In practice larger households have lower emissions per head across nearly all categories.
8 The original version of this uncomfortable epigram was resonant enough to appear in all three synoptic Gospels: Matthew XIX 24; Mark X 25, Luke XVIII 25.
9 The parallels with the sale of ecclesiastical indulgences, rejection of which led to the Protestant Reformation, have been frequently noted. George Monbiot for one, makes a wonderfully scathing modern Luther (Monbiot, 2006). See also the hilarious satire on offsetting found at www.cheatneutral.com <http//:www.cheatneutral.com>

## References

AECB (2006) www.aecb.net

Boardman, Brenda *et al.* (2005) *The 40% House*, Environmental Change Institute, University of Oxford.

Boyle, G., Everett, B. and Ramage, J. (2003) *Energy Systems and Sustainability*, Oxford: University Press.

Car Plus (2006) www.carplus.org.uk

Cyberium (2006) http://www.cyberium.co.uk/branding.htm

Dawson, J. (2005) *Ecovillages*. Schumacher Briefing Papers, Bristol. See also www.gen.org

DEFRA (2004) *Scientific and technical aspects of climate change, including impacts and adaptation and associated costs*, Department for Food and Rural Affairs, London. www.defra.gov.uk/environment/climatechange/pdf/cc-science-0904.pdf

DEFRA (2006) *The Environment in Your Pocket*, London: Department for Environment, Food and Rural Affairs.

Dresner, S. and Ekins, P. (2004) *Economic Instruments for a Socially Neutral National Home Energy Efficiency Programme*, Policy Studies Institute, Research Discussion Paper No. 18.

Dresner, S. and Ekins, P. (2006) 'Economic Instruments to Improve Home Energy Efficiency without Negative Social Impacts', *Fiscal Studies* **27** (1), 47.

Energy Saving Trust (2004) *Rural Biomass Community Heating Case Study*. Energy Efficiency Best Practice in Housing Document No. CE91.

Eurometrex http://www.eurometrex.org/Docs/Meetings/nurnberg_2005/EN_Climate_Change_Discussion.pdf

Food Climate Research Network (2005) http://www.fcrn.org.uk/lifecycl.htm

Forster, P.M. de F., Shine, K.P. and Stuber, N. (2006) It is premature to include non-$CO_2$ aviation effects in emission-trading schemes. *Atmospheric Environment*, 40, 1117.

Francis, P. (2004) The Impact of UK Households on Environment Through Direct and Indirect Generation of Greenhouse Gases, *Economic Trends* 611, London: Office of National Statistics.

Harper, P. (1999) 'Techno-Anthropology in the Home', *Radical Statistics* No. 71, Summer.

Harper, P. (2006) 'Food and Carbon Emissions', *Clean Slate* No. 62.
Harper, P. and Todd, B. (2004) 'How Low Can a Country Go?' *Renew* No. 150, June.
Harvard School of Public Health, http://www.hsph.harvard.edu/nutrition-source/pyramids.html
Hawken, Paul, Amory Lovins and Hunter Lovins (1999) *Natural Capitalism.* London: Earthscan.
Hertwich, E. (2003) The seeds of sustainable consumption patterns. *Proceedings, 1st International Workshop on Sustainable Consumption,* Society for Non-Traditional Technology, Tokyo 19–20 May.
Hertwich, E. (2003) Consumption and the Rebound Effect. Proc. SETAC-ISIE Case Study Symposium, Lausanne.
Hillman, Mayer (2004) *How We Can Save the Planet.* London: Penguin.
HM Government White Paper (2003) Our *Energy Future: Creating a Low-Carbon Economy.* London.
HM Treasury (2006) *The Stern Review: The Economics of Climate Change.* London.
http://www.dft.gov.uk/stellent/groups/dft_transstats/documents/downloadable/dft_transstats_035650.pdf
Institute for Fiscal Studies (2006) An online calculator for household incomes can be found on http://www.ifs.org.uk/wheredoyoufitin/.
IPCC *Climate Change 2001* (2001). Synthesis Report. Cambridge: CUP.
IPCC (2007) http://ipcc-wg1.ucar.edu/
Jackson, T. (2006) 'Beyond the "Wellbeing Paradox": wellbeing, consumption growth and sustainability.' Concept paper prepared for the New Economics Foundation as input to DEFRA Whitehall Wellbeing Working Group.
Jackson, T. (2005) 'Live better by consuming less? Is there a double dividend in sustainable consumption?' *Journal of Industrial Ecology* 9(1–2) 19.
Kaufman, R.K. (2004) The mechanisms for autonomous energy efficiency increases: A cointegration analysis of the US energy/GDP ratio. *Energy J.*, 25 (1).
Kauppi, P. and Sedjo, R. (2001) Technical and economic potential of options to enhance, maintain and manage biological carbon reservoirs and geo-engineering. In *Climate Change 2001: Mitigation. Contribution of Working Group III to the Third Assessment Report of the IPCC.* Cambridge, CUP.
Korbetis, M, Reay, D. and Grace, J. (2006) New Directions: Rich in $CO_2$, *Atmospheric Environment* **40**, 3219–220.
Layard, R. (2005) *Happiness: Lessons from a New Science.* London: Allen Lane.
Lee, J. (2003) 'The Potential offered by aircraft and engine technologies', in Upham, P. *et al.* (eds) *Towards Sustainable Aviation.* London: Earthscan.
Lorek, S. and Spangenberg, J.H. (2001) *Environmentally Sustainable Household Consumption.* Wuppertal Papers No. 117, Wuppertal Institute.
Matthew XIX 24; Mark X 25, Luke XVIII, 25.
Meltzer, G. (2005) *Sustainable Community: Learning from the Cohousing Model.* Canada: Trafford Publishing.
Meyer, A. and Hildyard, N. Climate and Equity after Kyoto. http://www.thecorner-house.org.uk/item.shtml?x=51957
Monbiot, G. (2006) *Heat: How to stop the planet burning.* London: Allen Lane.
New Economics Foundation (2005) *A Wellbeing Manifesto.* London, New Economics Foundation.

Noorman, K.J. and Uiterkamp, T.S. (eds) (1998) *Green Households?* London: Earthscan.

Passivhaus (2005) www.passivhaus.co.uk

Pretty, J. (ed.) (2005) *The Earthscan Reader in Sustainable Agriculture,* London: Earthscan.

Reinders, L. and Soret, S. (2003) Quantification of the environmental impact of different dietary protein choices. *Am. J. Chem. Nutrition,* **78** (3).

RMI (2006) see www.rmi.org

Royal Commission on Environmental Pollution, 22nd Report (2000) *Energy: The Changing Climate.* London 2000.

Royal Society (2002) *Economic Instruments for the Reduction of Carbon Emissions.* London.

Sausen, R. and others (2005) Aviation Radiative Forcing in 2000: An Update on IPCC 1999. *Meteorologische Zeitschrift,* **14** (4).

Sloman, L. (2006) *Car Sick.* Totnes: Green Books.

Sutherland, H., Taylor, R. and Gomulka, J. (2001) Combining household income and expenditure data in policy simulations. Microsimulation Unit Discussion Paper, Dept of Applied Economics, University of Cambridge.

Swingland, I.R. (ed.) (2002) *Capturing Carbon and Conserving Biodiversity.* London: Earthscan.

UK Department for Transport (2005) http://www.dft.gov.uk/stellent/groups/dft_transstats/documents/downloadable/dft_transstats_035650.pdf

UNEP (2003) The *Role of Product Service Systems in a Sustainable Society,* Paris, UNEP.

Upham, Paul *et al.* (eds) (2003) *Towards Sustainable Aviation,* London: Earthscan.

Vringer, K., and Blok, K. (1995) The direct and indirect energy requirements of households in the Netherlands, *Energy Policy,* Vol. 23, no. 10, pp. 893–910.

# 13
# A Sustainable Future for Energy?

*David Elliott*

This book has attempted to review the potential and problems of moving towards a long-term fully sustainable energy future. The assumption has been that the transition would be based initially on the adoption of more efficient generation and use of fossil fuels, and the sequestration of carbon emissions, in parallel with the expanding use of renewables, as a preliminary to a full transition to renewables.

It is clear that, in the interim, there is a range of ways in which the continued combustion of fossil fuels might be made more sustainable, including the adoption of Combined Heat and Power, and of carbon capture and storage techniques. While Chapter 6 looked in some detail at what can be done to improve the efficiency of energy use, we have not in this book focused on reducing emissions from the combustion of fossil fuels, in part because this area has been well covered elsewhere, even if sometimes a little optimistically (Jaccard, 2006). Instead, the emphasis has been on the later phase, i.e. mainly on the newly emerging renewable energy technologies, and it seems clear that these technologies could provide a viable basis for the longer-term future.

A shift in this direction will however require a new approach to energy generation and use, with more emphasis on decentralised energy generation coupled with better end-use energy management. That implies a shift away from traditional approaches, and, in particular, a move away from large-scale nuclear plants, an option which in any case, as argued in Chapter 1, has many short- and long-term problems.

The Sustainable Development Commission has argued the case for a more decentralised approach point well, focusing on the UK context: '*Reliance on centralised supply may exacerbate the current institutional bias towards large-scale generation, and the reluctance to really embrace the reforms necessary to ensure a more decentralised and sustainable energy economy'*. It added *'the lack of flexibility, or "lock-in", associated with investment in large-scale centralised supply like nuclear power is also a concern. This relates to the issue of sunk costs. A new nuclear programme would commit the UK to that technology, and a centralised supply infrastructure, for at least 50 years. During this time there are likely to be significant advances in decentralised technologies, and there is a risk that continued dependence on more centralised supplies may lock out some alternatives. Decentralised supply is generally more flexible because it is modular, and can adapt quicker and at less cost to changed circumstances. More locally-based energy provision may also be conducive to the sustainable communities agenda, a key part of the UK Government's Sustainable Development Strategy'* (SDC, 2006).

Certainly there is increasing interest in smaller scale, more decentralised approach to energy supply, in terms of for example community-scale generation. The Borough of Woking has been a pioneer in this regard. Its decentral generation projects, coupled with a range of energy efficiency projects, have allowed the Borough to reduced energy use by half and $CO_2$ emissions by over 77% between 1991 and 2004. A 'private wire' electricity supply network has been established, together with a heat distribution network, with heat and power being fed from a series of local CHP units. As Everett notes in Chapter 6, a fuel cell, providing both heat and power, is also linked in, as are a series of PV solar arrays on local buildings, providing electricity. Similar decentral energy programmes are now being planned across London, overseen by the London Climate Change Agency, with CHP seen as being able to pay a major role, some of it being biomass fired (PB Power, 2006).

In parallel there has also been increasing interest in domestic-scale micro generation, and some optimistic projections have been made of the potential. For example the Energy Saving Trust has suggested that by 2050 micro-generation units like micro-wind turbines, roof top PV solar and micro-CHP might be supplying up to 40% of UK domestic heat requirements and 25% of domestic electricity (EST, 2006). However, although there has been a lot of enthusiasm for micro-power, including from consumers keen to 'do their bit', some micro-power technologies are relatively new and untested on a wide scale. For example there have been doubts expressed about the practical performance of micro-wind turbines, especially in low wind urban environments (Martin, 2005),and the results of some of the early consumer trials with domestic micro-CHP Stirling engine units have not been too impressive (Carbon Trust, 2006).

While it will take time to develop new systems like this fully and to find appropriate ways to integrate them into the wider energy system, micro power clearly has its attractions, for example in terms of avoiding power losses from long distance transmission. In addition micro-power does provide a way to engage individual consumers in the transition to decentralised power, and this involvement can have positive collateral effects. For example, a study of consumer reactions to micro-generation for the UK Sustainable Consumption Roundtable found that 'beyond the sheer excitement and pleasure of DIY energy generation, the impact is seen in householders' shifting attitudes to energy conservation and consumption' (Hub, 2006).

One of the early visions of 'alternative technology' was that individual households could make a major contribution to self sufficiency by adopting small-scale renewable systems (Boyle and Harper, 1976). While fully 'off grid' systems are unlikely to be very relevant in many locations in the UK, the idea of self-generation using micro-power devices still motivates those consumers who are keen to limit the amount of power they buy in from commercial vendors. Community-scaled technologies also have their social attractions in providing a degree of local economic autonomy and control, and although the vision of substantially decentralised local self-management of energy systems is somewhat less apparent these days, there are some interesting examples of sustainable community projects, such as the Hockerton Housing Project in Nottinghamshire. In addition, as Toke describes in Chapter 8, there are also now examples of community-owned wind co-ops in the UK. However, projects like this are usually relatively small, and as Toke admits, there is also a role for corporate projects, and these are usually larger.

Certainly, while the general thrust may be towards smaller-scale projects, decentralisation, and localisation where possible, it is a matter of degree. There is also likely to be a need to use alternative supply technologies that are larger, and usually also more remotely located. Some offshore wind, wave and tidal current farms may be up to 1 GW generation capacity, and by their nature will mostly be away from centres of population. They are also usually corporately owned, although there may be opportunities for local community involvement. There may also be scope for some local control or even ownership, for example by municipal authorities, for some larger project, like tidal lagoon and barrage projects, whose location and scale is geographically determined. Some interim 'transitional' fossil-fuel technologies will also be large, and are likely to be corporately owned – for example Clean

Coal/CCS plants, which are also likely to have to be located near the coast to provide easy access to geological storage sites. However there can be a variety of sizes and locations for gas-fired CHP plants, and ownership can range from consumer-owned domestic micro-CHP units, to larger municipal CHP programmes in urban areas; as was demonstrated in the case of Woking, they can be managed by joint public-private Energy Services Companies. Longer term, as and when biomass-powered CHP begins to be used more widely, getting access to fuel will have locational implications, but even so it does seem likely that a range of scales and management structures will be possible, including, for example, biomass growing co-ops.

## The UK sustainable energy mix

Whatever the scale, location or ownership, there is plenty of room for debate when it comes to deciding on the best mix of sustainable generation technologies, and which should be developed first. For example, so far the emphasis tends to have been on electricity, rather than heat production. And yet many renewables can supply heat direct – notably solar and biomass. As was illustrated in Chapter 5, the solar heating option has much to commend it – it can supply heat direct to where it is needed on a range of scales. But biomass has the major attraction that it can be stored. The UK Biomass Task Force suggested that biomass might provide up to 7% of UK heat requirements by 2015.

However, there is much debate about the best role for biomass. Some think that biomass crops and wastes should be used mainly for electricity generation, rather than for direct heating, while some look to biogas production for heating as less environmentally damaging. Others say that it will be easier to go for biofuels for vehicles – an easier, more direct market. Certainly as Scurlock argues in Chapter 3, there is a significant potential, particularly for the so-called third generation biofuels.

Then again, the hydrogen lobby sees the use of biofuels as at best preliminary for the use of hydrogen as fuel for vehicles. Indeed some see hydrogen as becoming a major general purpose secondary fuel for many end uses. It could be sent down a pipe from central production units, admixed with, or eventually replacing, natural gas, and used directly for heating, or for local electric power production in a fuel cell. In addition, it could be used as fuel for vehicles either directly, by combustion, or to generate electricity in an on-board fuel cell. Initially the hydrogen might be generated by conventional steam reformation using fossil fuels (sometimes then being termed 'brown' hydrogen'), but it could increasingly come directly or indirectly from renewables (then being called 'green hydrogen'). It can be produced by direct biomass and waste conversion (e.g. using advanced gasification and pyrolysis techniques), and also by the electrolysis of water, using electricity from wind, wave and tidal sources. Some of course also see nuclear power playing a role in this context, perhaps even producing hydrogen by high temperature dissociation of water (so-called 'black hydrogen'), although that seems a long shot. High temperature focused solar systems might be a better bet.

Although there are obvious issues with safety, as well as with the energy losses associated with electrolysis and fuel cell conversion, these do not seem insurmountable – for example the waste heat from conversion processes can be captured and used; and hydrogen is already used safely as fuel for spacecraft launchers and in some road vehicles. Certainly there is a lot of enthusiasm for a shift to a renewables-powered hydrogen future (McAlister, 2003). The hydrogen option clearly has specific attractions in relation to renewables, since it is a storable fuel, providing a way to compensate for the variability and intermittency of some renewable energy sources. Although as Milborrow makes clear in Chapter 2, the variability of wind and other renewables is not a major problem at low levels of penetration in the energy system, it could be crucial for the longer term, as the proportion of variable renewables in the overall energy mix increases.

However, there is plenty of room for disagreement over when a move to a hydrogen economy based on renewables might begin, with the hydrogen being used for multiple roles – for electricity, heat and transport. Currently the main focus, particularly in the US, is on ways to generate hydrogen from fossil fuels for use as vehicle fuel, possibly using intermediate fuels like methanol converted to hydrogen in on-board reformers. Many environmentalist fear that following up 'brown' hydrogen options like this could delay what they see as the crucial and urgent shift to 'green' hydrogen from renewables (Dunn, 2001).

Unfortunately though there could problems with trying to go straight to 'green' hydrogen to meet transport requirements- initially there may not be enough renewables. After all, wind, wave and tidal energy systems are usually seen, initially at least, as playing a central role just in *direct* electricity production, distributed by the power grid. Similarly PV solar is seen as supplying electricity direct where needed. It may be that conversion to hydrogen for interim storage is seen as useful in order to respond to the problem of the intermittency of some renewables, but the main final output would presumably be electricity. At least initially, there is unlikely to be enough renewable capacity to provide large-scale hydrogen production for other uses as well. Some might be used for heating, but that seems to be a poor use of an expensively produced fuel, when there are easier way to provide heating *via* renewables. While at some point it might be possible to use green hydrogen to run some types of vehicles (e.g urban buses) it can be argued that electricity obtained from valuable renewable energy sources should not be used to produce hydrogen just to keep private vehicles going, and certainly not in the short term. Similarly, valuable biomass, and land, should not be used to create hydrogen directly for this purpose. More generally, it is sometimes argued that the transport problem needs more than just technical fixes like this: they may work temporarily, but will allow us to avoid tackling the more fundamental issues of how to manage transport demand. On this view what is needed is social change rather than new technology. The reality though is probably that *both* approaches may be needed – social and technical change (Potter, 2006).

Although the transport issue has not been covered in this book in any detail, Chapter 7 did mention some potential social fixes. However, given the strength of demand for transport, and resistance to the significant behavioural adaptions needed to reduce it, it seems inevitable that technical fixes will be increasingly popular. It is sometimes argued that, one way to avoid using renewables to meet this demand might be, initially at least, to leave the production of hydrogen to the new generation of clean coal gasification plants, with the resultant carbon dioxide being captured and stored in undersea wells. Relying on 'brown' hydrogen might then be condoned as an interim option, leaving renewables free to meet direct electricity, and, where relevant, heating requirements, until they were able, at a later stage, to also play a role in supplying hydrogen for the transport sector.

Overall, given that hydrogen systems are still in their infancy, the most likely initial approach to the development of sustainable energy in the UK context seems to be a continued focus on direct electricity generation, with increasing reliance on wind. There are limits to how much capacity can be installed on land, and there will be transmission grid strengthening issues to face, especially as Scotland's very large on-land wind resource is developed. At present it is unclear who will pay for the necessary grid upgrades, although it could be argued that the energy supply companies, who have done very well over the past decade or so from relatively high charges for power sold to consumers, ought to avoid the risk of government imposed wind-fall taxes by investing in grid infrastructure for the future. One idea for bringing the power from the North to the South, where most of the demand is, while avoiding the considerable planning permission problems of getting access across the land, is to install 1–2 GW capacity marine cable offshore. One possibility would be a High Voltage Direct Current cable on the sea-bed between the Chapelcross nuclear site in southern Scotland, which is currently being decommissioned, and the soon to close Wyfla nuclear plant on Anglesea in N Wales. Both sites already of course have links to the mainland grid. The cost would be perhaps £800 million.

However, once major undersea links are being considered, it begins to make sense to think in terms of exploiting the very large deep-sea offshore wind resource. The UK already has an offshore wind farm programme, with projects of up to 1 GW in shallow coastal waters, and this seems likely to expand, and go further and further offshore as experience is gained with deep sea installation. Moreover, at some point this might involve a north sea supergrid, as proposed by Airtricity, who have talked of an initial 10 GW 'foundation project' supplying power to the UK, the Netherlands and Germany (Airtricity, 2006).

Wind power is clearly the current front runner for electricity production, but in addition, as was argued in Chapter 4, wave power (on shore, inshore and increasingly deep sea) and tidal power (tidal current turbines, lagoons and maybe even barrages) should also be able to play an increasing role. At some point projects like this may provide sufficient power to support some hydrogen production, which might lessen the problem of providing grid links from relatively remote locations. It has even been suggested that hydrogen can be generated onboard floating wind, wave of tidal rigs out at sea, by the electrolysis of sea water, and then tanked back to shore by ships.

Meanwhile, increasing use could be made of biomass of various types for a variety of end uses, including heat as well as possibly hydrogen production, and electricity production, although there are obvious land-use implications to how much energy can be made available in the UK from biomass. By contrast, solar power, for both heating and electricity production, is obviously less constrained in terms of space, since it can be roof mounted. While it may be some while before it will be economically viable, in the UK context, to use PV solar for hydrogen production, PV solar seems likely to become increasingly popular for direct electricity production, particularly in day-time occupancy buildings, but also, if net metering systems develop, in the domestic sector.

Although not a renewable energy system as such, the use of electric powered ground-source heat pumps is another option that is likely to become increasingly popular, especially in areas where CHP/district heating links are not viable. And, with energy prices rising, geothermal energy might make a comeback for both heating and power production – the UK's £50 million aquifer and 'Hot Dry Rock' programme was abandoned in the late 1980s on economic grounds. Geothermal energy has the big attractions that, unlike wind, wave, and solar, it is not weather related. Finally, valuable gains can be made in some locations from micro and mini hydro, while some of the UK's large-scale existing hydro plants can be upgraded – they are now about the cheapest source of power on the grid.

Overall, it does seem that renewables could make a very significant contribution by say 2050. Figure 13.1 is a recent scenario by George Marsh from Future Energy Solutions, suggesting that renewables plus Combined Heat and Power could supply around 50% of UK electricity

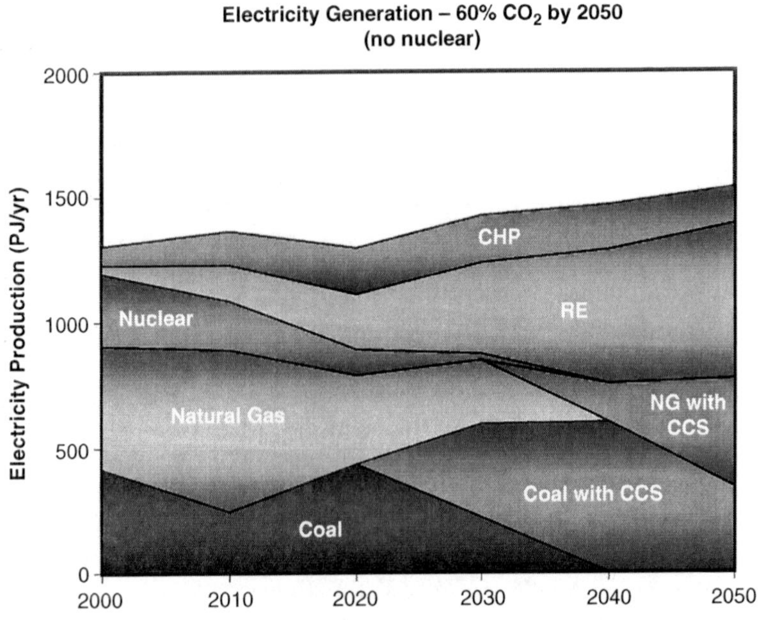


*Figure 13.1* A UK electricity projection from Future Energy Solutions, with renewables and CHP proving 50% by 2050 and coal and the remainder met by gas and coal plants with Carbon Capture and Storage: emissions fall by 60% by 2050

***Table 13.1*** **Potential % of overall UK electricity supply in 2050**

| | |
|---|---|
| Onshore wind | 8–11% |
| Offshore wind | 18–23% |
| Wave/Tidal | 12–14% |
| Biomass | 9–11% |
| PV solar | 6–8% |
| TOTAL | 53–67% |

Based on overall likely level of supply of 400–500 TWh in 2050

*Source*: DTI/Carbon Trust 'Renewables Innovation Review' 2004.

by then. In this scenario, demand is held down by a major commitment to energy efficiency, and the continued combustion of gas and coal makes up the remaining 50%, with the associated carbon dioxide emissions sequestrated *via* capture and storage technology (Marsh, 2005). Moreover it could be that renewables could do better than this. Table 13.1 shows an assessment of overall UK potential for some of the key options by 2050, based on the 2004 Renewable Innovation Review carried out by the DTI/Carbon Trust. Some of these projections might be seen as pessimistic. For example, as was argued in Chapter 4, by 2050 wave and tidal power ought be able to do much better than indicated in Table 13.1. So might many of the others. In this context it is worth noting that by 2007, Scotland was obtaining 18% of its electricity from renewable sources, while according to Scottish Renewables, by 2020 more than 50% could come from renewables (Scottish Renewables, 2006). Scotland may be particularly blessed with renewable resources, but even the prospect, indicated by Table 13.1, of the UK as a whole being able to obtain around 50% electricity from renewables by 2050 does provide a good basis for a discussion of targets and strategy – for example with regards to which renewable energy options should be developed first and how to create a sensible balance.

## Choosing amongst the options

In trying to decide which technologies to promote, and when and how much to promote them, economic assessments will inevitably be central. Although it is hard to project costs for relatively new technologies decades ahead, the studies by the Cabinet Office Performance and Innovation Unit in 2002 gave some indication of possible price/kWh trends, based on learning curve extrapolations (see Table 1.5). It suggested that wind power would come out to be very attractive by 2020 – at 1.5–2.5 p/kWh on land and 2–3 p/kWh offshore. On this basis it seems sensible to press ahead with wind, particularly offshore, which has the largest UK renewables electricity potential. In addition, although generation prices may be higher, a strong case can also be made for wave and tidal, which are well suited to the UK in terms of its technological capabilities, the marine energy resource potential and the technology export opportunities. For example in a report reviewing the most cost-effective options for reducing carbon emissions from a commercial 'UKplc' perspective, the Carbon Trust noted that. 'Early estimates show that the UK has very significant wave and tidal stream resource, a strong competitive position and, therefore, the potential to become a global leader in the medium term if these technologies can become cost competitive' (Carbon Trust, 2003).

While playing to your strengths clearly makes sense, these benefits tend to be longer term, which implies a key role for government, which can take a longer-term strategic view than the private sector. In 2003, the House of Commons Select Committee on Science and Technology argued that 'the Government has an important role in identifying those of Britain's strengths that are consistent with the industrial environment and the market. It should provide a clear and unambiguous focus' (Select Committee, 2003).

However this strategic approach to choosing technologies has its problems. For example some argue that governments are poor at making such choices, and that the choice of technology should be left to competitive mechanisms and market forces, presumably as interpreted by private corporate/investment interests. In his influential review of the Economics of Climate Change and strategies for responding to it, Sir Nicholas Stern warns against government officials trying to 'pick winners', but then argues that, 'technology-neutral incentives should be complemented by focused incentives to bring forward a portfolio of technologies' (Stern, 2007).

He suggests that this focused approach is necessary since 'the process of learning means that longer-established technologies will tend to have a price advantage over newer technologies, and untargeted support will favour these more developed technologies and bring them still further down the learning curve'. He goes on 'This concentration on near-to-market technologies will tend to work to the exclusion of other promising technologies, which means that only a very narrow portfolio of technologies will be supported'.

In order to take a wider, longer-term view, governments therefore need to make strategic selections amongst the technologies, and as noted in Chapter 9, Stern backs the technology-specific Feed-In Tariff approach against the market-orientated technology-neutral Renewables Obligation, as well as a major expansion of Research and Development funding – he suggest 2–5 times more. Although he suggests that there is also a need for an 'escape route', with 'a clear review process and exit strategies', he adds 'governments must accept that some technologies will fail. Uncertainty over the economies of scale and learning-by-doing means that some technological failures are inevitable. Technological failures can still create valuable knowledge, and the closing of technological avenues narrows the investment options and increases confidence in other technologies'.

The selection – and deletion – of technologies is thus not a static, once only, activity, it is a ongoing process, which also has to take account of new emerging external factors. While there is clearly a need to press ahead with developing portfolios of technological options and appropriate support schemes, it also must be recognised that the technological, economic and social context is changing, and there is a need to learn as we go along. The best long-term overall balance between the various options and scales will probably only emerge as the sustainable energy system develops, and as more experience is gained with the new systems and associated infrastructure. In particular, what finally emerges is likely to depend not only on such factors as the extent of the development of hydrogen as an energy carrier, but equally importantly, on the degree of success in getting demand under control. That is not just a technical issue.

Given the problems of 'rebound', as outlined by Herring in Chapter 7, and the seemingly inexorable rise of consumer expectations, getting demand under control will presumably involve and require more than the introduction of simple 'technical' efficiency measures, important though they are. It will also need system-level 'demand management' techniques like interactive load management, smart metering and, as Everett notes in Chapter 6, the development of an 'informated grid'. However, in addition to these technical changes, getting demand under control also seem likely to involve and require lifestyle changes, and changes in consumption patterns. That opens up the wider issue of social acceptance of change.

## Choosing lifestyles

For some people, one attraction of renewable energy is that it seems to offer a way to respond to environmental problems like climate change without having to make painful changes in lifestyles and society generally. Some others believe that a sustainable future will require radical social and economic change, changes which are seen as either a threat, or as desirable, depending essentially on the political perspective adopted. There is certainly no shortage of challenging social and political issues in that context, for example the question of whether voluntary change will be sufficient or whether change will have to be imposed or at least strongly stimulated.

In Chapter 12, Peter Harper looked at what individuals might do voluntarily to reduce their carbon footprints, and in Chapter 7 Herring also provides some examples of positive 'grassroots' self-help initiatives. However, as Herring notes, there are also proposals for more draconian measures, for example personal carbon credit 'rationing' systems, with an opportunity for buying up/trading in unused credits. While carbon rationing/trading would provide a powerful way of educating people about their energy use, especially when and if the carbon cap allocation levels were progressively reduced, it could be socially divisive and economically regressive, and even counter-productive in carbon terms. For example, quite apart from the risk of corruption and falsification, it would enable the rich to continue to use energy unconstrained, by buying carbon credits from the poor, who might be tempted to sell all of their increasingly valuable credits and find ways to buy in 'black market' high-carbon energy. There may be ways to avoid this type of outcome e.g. *via* tighter caps, regulation and policing, but they too have social implications.

Another less invasive approach might be to rely on voluntary carbon offsets – consumer-orientated 'dispensation' systems. Interestingly Harper, in Chapter 12, seems willing to condone carbon offsets as 'better than nothing', while Peake, in Chapter 10, felt that some at least were unverifiable and even unethical – in effect dumping the problem on someone else and allowing the rich to buy their way out of responsibility for their emissions.

Whatever ones views on the specifics of offset or rationing schemes, it is clear that energy and climate policy cannot ignore social and economic divisions and imbalances- something that is also been central to campaigns for the reduction of fuel poverty. This opens up wider issues of social change. To put it simply, is it to be taken as given that there will always be a group of poor and a group of rich people, with different needs and opportunities – in each country and across the world? Or might not at least part of the answer to climate change be a reduction in these inequalities?

However, that issue takes us well beyond the reach of this text. Rather than exploring the wider social and political issues, what this book has tried to explore is, primarily, whether the various sustainable energy options are viable technically and economically, focusing on the technology and associated systems. The message seems to be that, although there are many challenges, these technical options do seem to offer a viable future, although new ways of supporting their rapid development will be needed. Whether that will be sufficient to deal with future rates of economic growth is of course less clear, with the ever increasing demand for transport being perhaps the most challenging issue. But it does seem clear that, whatever else is done (for example in terms of moving to a more sustainable approach to consumption), the contribution from renewables should be increased as rapidly as possible. The issue then is how best to move in that direction and which technologies to support in order the make the transition.

## The global perspective

For this sort of transition to have a significant impact on climate change, it has to occur worldwide. Overall, at the global level, while the problems are clear, there do also seem to be major opportunities, for the development of sustainable energy, as the scenarios discussed by Boyle in Chapter 11 suggest. Certainly, as emerged from the International 'Renewable 2004' Conference organised by the German Federal Government in 2004 in Bonn, it does not seem utopian to suggest that it should be possible to obtain 50% of total world energy from renewables by 2050 and of course more after that.

The main focus in this book has been on possible developments in industrialised countries like the UK, where there are well established national grids and a range of existing fossil and, in some cases, nuclear power plants. So the transitional programme in that context initially involves adjustments which reduce emissions while renewables are deployed- and which also lay the basis for the transition. For example, providing a heat grid for fossil-fuelled CHP plants creates a valuable infrastructure which can subsequently be used when heat from renewables like biomass becomes widely available. Similarly, the development of technologies for the generation, transmission, storage and use of hydrogen gas, produced initially from fossil fuels, can lay the basis for a more sustainable system using renewable energy to provide the power to generate hydrogen for a wide range of uses.

In much of the rest of the world, the situation is usually very different. Although energy demand is nowhere near as large, power grids are limited or even absent. Traditional biomass often plays a major role, along with small and often very inefficient fossil fuel plants, plus a few small renewable energy projects – micro hydro, solar and sometime wind and biogas. For those countries that are currently rapidly industrialising, there is the temptation to simply copy the west and invest in fossil or even nuclear plants. Clearly there are major problems associated with rapid economic development, not least the sudden increase in energy demand, which may make it hard to avoid using fossil fuels for an interim period. In which case, adopting the cleanest technology, including carbon capture and storage, will be important. However, for some, starting as it were from scratch, there may also be an opportunity to 'leapfrog' over the old technology and old fuels, and develop a fully sustainable energy system, especially if financial and technical aid can be provided.

Various approaches have been suggested for supporting the development of sustainable energy technologies such as renewables, and experience has been gained with some of them. Chapter 8 looked at the role of local ownership and in many ways this seems to be a win-win option, creating both local support and local benefits. However there may be limits to what can be achieved in this way, especially when it comes to newer technologies. Chapter 9 looked at two of the main conventional contenders for supporting newly emerging or near-market options at the national level – feed-in tariffs and quota/trading schemes. It was concluded that, so far, REFIT Feed-In Tariff type schemes were clearly more effective, but longer term, as the technologies and markets matured, green certificate/emission trading schemes were probably best, as a follow-up approach. As Chapter 10 illustrated, trading schemes also seem likely to be increasingly relevant at the international level. Indeed, market mechanisms of this sort, set within a context of mandatory emission caps defined by negotiation, are fundamental to the Kyoto approach.

However there is an alternative or perhaps additional approach, which has been promoted by the Asian-Pacific Partnership on Clean Development and Climate Change – backed by the US, China, India, Japan, S. Korea and Australia – which avoids mandatory targets and certificate trading. Instead the emphasis is simply on technology, the aim being to stimulate and co-ordinate commercial transfer of relevant technologies from those countries that have developed them to with those that need them. While they agree that technology transfer in important, supporters of the Kyoto approach however argue that transfer is only likely to happen on a significant scale in the context of an international regime with mandatory targets, and some fear that the Asian-Pacific Pact will in fact undermine the Kyoto approach. The counter argument is that the transfer arrangements established within the Kyoto framework (*via* the Clean Development Mechanism) are bureaucratic and slow.

It is certainly true that countries like the US have enormous capacity for developing the new technologies needed to reduce emissions, but it remains unclear whether a purely technology-led approach, within a basically unchanged global market context, will suffice. Political intervention, and the establishment and operation of an appropriate regulatory framework, seems to be necessary to create the economic conditions for the rapid uptake of the new technologies and development of the necessary institutional and social infrastructure of support and acceptance.

## Transition politics

Making this transition will not be easy, but there are some guides to how it might be managed. In addition to practical experience with introducing the new energy technologies, over the past few years a body of analysis of 'transition theory' has emerged, drawing on the work of Geels *et al.* in the Netherlands (Geels, 2004). This has been developed into a conceptual framework that has been applied to the UK energy system by, amongst others, the Tyndall Centre (Anderson *et al.*, 2005). Parallel developments in innovation theory are also relevant – highlighting the importance of 'bottom-up' processes, often based on ideas developed initially at the grass roots for niche markets, as opposed to the traditional 'top-down' approach (Douthwaite, 2002). That is how wind power got started so successfully in Denmark, and 'strategic niche management' is now seen as an important tool for identifying and then promoting the initial development and then successful wide-spread diffusion of new technologies, including renewable energy technologies (Smith, 2005).

One emphasis in the new approach to technological transition is an awareness of the differing needs for each phase of the innovation process, an insight reinforced by experience with learning curves, which illustrate how costs and performance can improve with volume production and market diffusion (Elliott, 2006). The message is that new technologies need the right kind of support at each stage and a changing but continuing framework of support as they develop, if they are to help create a full-scale transition in economic, technological and social terms. Ad-hoc programmes (stop-go), overly ambitious targets (like the 4 GW wave device target set in the UK the 1970s) and lack of follow-through for emerging technologies (again as evidenced by the early UK wave energy programme), have to be replaced by flexible but consistent support programmes, usually with an emphasis on building on and expanding out of initially protected niches.

As the niche concept suggests, an underlying insight is that purely technological development are not enough. The technology has to be widely adopted and that means social as well as market acceptance. It may also mean social initiation – which is one reason why grass-roots initiatives and locally owned community projects looked at in Chapter 8 are important. They can provide the niche context for developing new ideas about how technology can be used, and also about how lifestyles might be adjusted. The point is that interaction between technology and society must be two way: it is not simply a matter of technology shaping society, or even of vice versa. Technology and society both change each other. Which means that social changes *and* technological changes must be expected. As noted earlier, not everyone will welcome social change, depending on its scale and who it impacts on. So continuing battles over development strategies and policies can be expected. And beyond that is the larger question of whether quite *radical* social changes may be required for a truly sustainable future.

While this book has focused on the technical aspects, it has not ignored the wider issues. It seems likely that, as Harper argues in Chapter 12, there will be a need for more than just technological change and minor social adjustments. Some say the stage has already been reached when significant lifestyle changes, particularly for the affluent minority, are urgent. Indeed some argue that without radical changes in how and what is consumed, there will be no long-term future for humanity. Others argue that, if properly developed, the global renewable resource is large enough to provide for all our needs into the foreseeable future. The amount of energy available from the incident solar energy that drives most of the earth's natural energy flows is certainly large, but there are inevitably limits to how much energy can be harvested effectively and safely from these flows, and delivered efficiently to where it is needed. Moreover, there are other limits which may emerge before the ultimate energy-availability constraints are reached, the most obvious being land-use constraints. In addition, although the energy is in effect free, all energy conversion technologies need material resources, which are finite.

Overall, it seems unlikely that, even using renewable energy sources, economic growth of the type that is dominant at present can continue indefinitely on a planet with inevitably limited material resources. Although use can be made of renewable energy flows and ways can be found to 'do more with less' in terms of resources, there must be ultimate limits to the planets ability to meet humanities continually expanding material expectations. Technology can only take us so far. In the final analysis, there may be a need to learn to adjust expectations and develop ways of living which satisfy needs without further undermining the ecological viability of the activities of human beings. Moreover, if the threat of climate change is as serious as it now seems, then, in addition to making radical technological changes, there may be a need to face up to the need to move to a more sustainable approach to consumption earlier rather than later.

Discussion of the idea of 'sustainable consumption' has only just got seriously underway in the UK. Some of the issue involved were introduced in Chapters 7 and 12 and are taken up in more detail in another book in this series – 'Energy Efficiency and Sustainable Consumption: Time to deal with the Rebound Effect'. Tackling our addiction to ever increasing material consumption, and the role of consumption in our economic and social systems, is certainly a challenging problem. As Cohen *et al.* have noted that *'any action that tries to limit the use of material objects but does not offer alternative ways of satisfying social and psychological objectives is likely to fail'* (Cohen *et al.*, 2001).

However, whatever is done about consumption, the rapid development of renewables also seems an urgent priority. Hopefully this book will provided an indication of some useful directions ahead.

## References

Airtricity (2006) 'European Offshore Supergrid proposal: Vision and Executive Summary', Airtricity, London.

Anderson, K., Shackley, S., Mander, M. and Bows, A. (2005) 'Decarbonising the UK' Tyndall Centre report, Manchester University. See also Shackley, S. and Green, K. (2007) 'A conceptual framework for exploring transitions to decarbonised energy systems in the United Kingdom', *Energy*, 32, pp. 221–36.

Boyle, G. and Harper, P. (1976) 'Radical Technology', Wildwood House, London.

Carbon Trust (2003) 'Building options for UK renewable energy', Carbon Trust report, London.

Carbon Trust (2005) The Carbon Trust's Small-Scale CHP field trial update, Carbon Trust, London, November.

Cohen, M. and Murphy, J. (eds) (2001) *Exploring Sustainable Consumption: Environmental Policy and the Social Sciences*. Pergammon, London.

Douthwaite, B. (2002) 'Enabling Innovation', Zed Publications, London.

EST (2006) 'Potential for Microgeneration, study and analysis', Energy Saving Trust, London.

Dunn, S. (2001) 'Hydrogen Futures: Toward a Sustainable Energy System', Worldwatch paper 157, Worldwatch Institute, Washington DC.

Elliott, D. (2006) 'Diffusion: Consumers and Innovation' Block 4 T307: 'Innovation: Designing for a Sustainable Future', Open University, Milton Keynes.

Geels, F. (2004) 'Understanding system innovations: A critical literature review and a conceptual synthesis'. In: Elzen, B., Geels, F., Green, K. (eds) 'System innovation and the transition to sustainability: theory, evidence and policy', Edward Elgar, Cheltenham.

Hub (2006) 'Seeing the light: the impact of micro-generation on the way we use energy', Hub report for the Sustainable Consumption Roundtable, National Consumer Council and the Sustainable Development Commission, London.

Jaccard, M. (2006) 'Sustainable Fossil Fuels', Cambridge University Press, Cambridge.

Marsh, G. (2005) Presentation on behalf of Future Energy Solutions to Parliamentary Renewable and Sustainable Energy Group, Westminster, January 26th.

Martin, N. (2005) Micro wind power review in *Building for a Future*, Vol. 15, No. 3, Winter 2005/6. See also Sunday Times 25/6/06 and Hunt, M. (2007) 'Micro-wind reality check', *Renew* 166, March–April.

McAlister, R. (2003) 'The Solar Hydrogen Civilisation', American Hydrogen Association, Mesa AZ.

PB Power (2006) 'Powering London into the 21st century', report by PB Power for the Mayor of London and Greenpeace, London.

Potter, S. (2006) 'Travelling Light: The roles of behavioural change and technical innovation in achieving sustainable transport', Keynote paper for the 2006 Shell Eco-marathon.

SDC (2006) 'The role of nuclear power in a low carbon economy', SDC position paper, Sustainable Development Commission, London.

Select Committee (2003) 'Towards a Non-Carbon Fuel Economy: Research, Development and Demonstration'. House of Commons Select Committee on Science and Technology, Session 2002–3 Fourth Report, London.

Scottish Renewables (2006) 'Delivering the New Generation of Energy: A Route Map to Scotland's Renewable Energy', see: www.scottishrenewables.com/data/reports/

Smith, A. (2005) 'Supporting and harnessing diversity? Experiments in Alternative Technology', Final Research Report to the Economic and Social Research Council on a research project funded under the ESRC Sustainable Technologies Programme (2003–2005) www.sussex.ac.uk/spru/environment/at

Stern, N. (2007) 'The Economics of Climate Change', Review for HM Treasury by Sir Nicholas Stern, Cambridge University Press. http://www.hm-treasury.gov.uk/independent_reviews/stern_review_economics_climate_change/sternreview_index.cfm

# Index